The Scientific Basis for Vitamin Intake in Human Nutrition

Bibliotheca Nutritio et Dieta

No. 52

Series Editors *P. Walter,* Basel
J.C. Somogyi, Rüschlikon-Zürich

Basel · Freiburg · Paris · London · New York ·
New Delhi · Bangkok · Singapore · Tokyo · Sydney

The Scientific Basis for Vitamin Intake in Human Nutrition

Volume Editor *P. Walter*, Basel

34 figures and 29 tables, 1995

Basel · Freiburg · Paris · London · New York · New Delhi · Bangkok · Singapore · Tokyo · Sydney

•••••••••••••••••••••••••

Bibliotheca Nutritio et Dieta

Library of Congress Cataloging-in-Publication Data
European Academy of Nutritional Sciences. Workshop.
The Scientific basis for vitamin intake in human nutrition / editor, P. Walter.
(Bibliotheca nutritio et dieta; no. 52)
"EANS Workshop, Cannes, May 1–3, 1994".
Includes bibliographical references and index.
1. Vitamins in human nutrition – Congresses. 2. Vitamins – Physiological effect – Congresses. I. Walter, Paul, 1933– . II. Title. III. Series.
TX341.B5 no. 52
[QP771]
613.2 s–dc20
[612.3'99]
ISBN 3–8055–6166–0 (alk. paper)

Drug Dosage. The authors and the publisher have exerted every effort to ensure that drug selection and dosage set forth in this text are in accord with current recommendations and practice at the time of publication. However, in view of ongoing research, changes in government regulations, and the constant flow of information relating to drug therapy and drug reactions, the reader is urged to check the package insert for each drug for any change in indications and dosage and for added warnings and precautions. This is particularly important when the recommended agent is a new and/or infrequently employed drug.

Printed in Switzerland on acid-free paper by Thür AG Offsetdruck, Pratteln
ISBN 3–8055–6166–0

Contents

Introduction and Objectives

Additional Physiological Functions of Vitamins

The Potential of Vitamins in the Prevention of Chronic Diseases

Vitamin Intake in Europe

General Information on Today's Recommendations for Vitamin Intake

Conclusions

Walter P (ed): The Scientific Basis for Vitamin Intake in Human Nutrition.
Bibl Nutr Dieta. Basel, Karger, 1995, No 52, pp 1–6

Do We Need New Concepts for Establishing Recommended Dietary Allowances?

Paul Walter

Swiss Vitamin Institute and Department of Biochemistry, University of Basel, Switzerland

In order to prevent vitamin deficiencies, it is necessary to know the daily intake for each of the vitamins concerned. Many countries have produced quantitative dietary recommendations under a variety of names. Most people refer to the so-called RDAs, i.e. recommended dietary allowances, and the generally accepted definition is as follows [1]:

'RDAs are the levels of intake of essential nutrients that, on the basis of scientific knowledge, are judged by the Food and Nutrition Board to be adequate to meet the known nutritional needs of practically all healthy persons.'

A major problem in formulating RDA values is the fact that nutrient requirements differ between individuals. Conventionally they are assumed to have a normal Gaussian distribution as shown in figure 1, with a peak at the mean requirement, a lower threshold at −2 SD and the RDA value at +2 SD. The RDA value therefore gives an intake that would cover the needs of 97.5% of the group for which the Gaussian curve has been established. This value, being a group value, is often misinterpreted by individuals. It is often regarded as the lowest acceptable intake which is, of course, not true since most of the individuals will actually meet their requirement at a lower value than the RDA.

The terminology for dietary standards has been changing even though the general definition given above is still valid. In table 1 the term 'LRNI' and 'LTI' refer to the lowest threshold intake, i.e. mean value −2 SD as shown

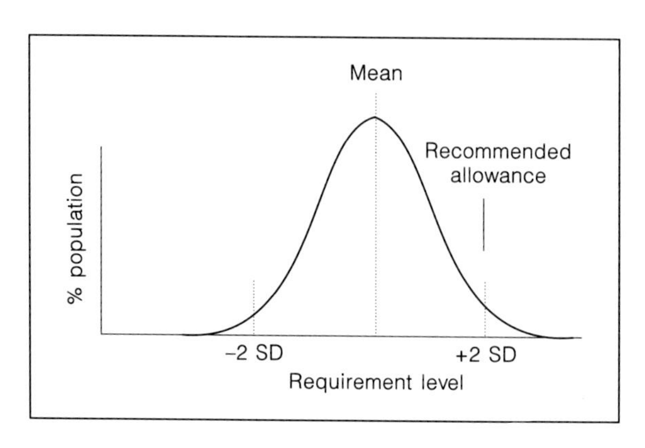

Fig. 1. Gaussian distribution of RDA values.

Table 1. Terminology for dietary standards

RDA	Recommended dietary allowances	USA (1941)
RDA	Recommended daily amounts	UK (1979)
RNI	Recommended nutrient intakes	Canada (1983)
RDI	Recommended daily intakes	Australia (1986)
LRNI	Low reference nutrient intake	UK (1991)
EAR	Estimated average requirement	
RNI	Reference nutrient intake	
LTI	Lowest threshold intake	EC-SCF (1993)
AR	Average requirement	
PRI	Population reference intake	

in figure 1. 'EAR' and 'AR' refer to the mean values, whereas 'RNI' and 'PRI' are identical with the 'RDA', i.e. mean value +2 SD.

When we take a look at the changes in the RDAs over the years, it is interesting to note that the RDAs of the National Research Council (NRC) in the United States have changed on the whole only little between 1968 and 1989 (table 2). Furthermore, the NRC RDAs of 1989 are very close to the 1990 values of the European Community (EC) (table 3) that are valid now. However, looking at more recent values from the United Kingdom of 1991 [2] and at those from the proposal of the Scientific Committee of Food (SCF) [3] of the EC of 1993, there is a strong tendency for lower RDA values or even for no values at all. In the case of pantothenic acid, biotin and vitamin D, only 'acceptable ranges of (safe) intakes' are given. In the case of vitamin

Table 2. History of the NRC RDAs for vitamins (adult males)

Vitamin	1941	1948	1957	1968	1976	1980	1989
Vitamin A, mg RE	1,000	1,000	1,000	1,000	1,000	1,000	1,000
Vitamin D				400 IU	400 IU	5 μg	5 μg
Vitamin E, IU				30	15	10	10
Vitamin K, μg							80
Vitamin C, mg	75	75	70	60	45	60	60
Thiamin, mg	2.3	1.5	0.9	1.3	1.4	1.4	1.5
Riboflavin, mg	3.3	1.8	1.3	1.7	1.6	1.6	1.7
Niacin, mg	23	15	15	17	18	18	19
Vitamin B_6, mg				2.0	2.0	2.2	2.0
Pantothenic acid							
Biotin							
Folate, μg				400	400	400	200
Vitamin B_{12}, μg				3.0	3.0	5.0	2.0

Table 3. RDAs for vitamins in Europe (adult males)

Vitamin	EC[1] 1990	UK 1991	EC-SCF 1993
A μg	800	700	700
D μg	5	–	–
E mg	10	–	–
C mg	60	40	45
B_1 mg	1.4	1	1.1
B_2 mg	1.6	1.3	1.5
Niacin mg	18	17	18
B_6 mg	2	1.4	1.5
Folate μg	200	200	200
B_{12} μg	1	1.5	1.4
Biotin μg	0.15	–	–
Pantothenic acid mg	6	–	–

[1] RDA values used for labelling (EEC council directive 90/496/EEC).

E the new SCF value is based on the number of double bonds in the estimated amount of polysaturated fatty acids (PUFA) in the diet.

The question can be asked: Do we need less vitamins? The answer is no, however, using the same criteria as in earlier years, better and more accurate evidence has become available for setting the new values. In addition, some of the 'safety additions' that had been added earlier, have now been reduced in many cases.

It is important to realize that also the new evaluations of the RDAs are still based essentially on the vitamin requirements to prevent the classical deficiencies. We know however, that already after relative short periods of unbalanced food intake, marginal vitamin deficiencies occur resulting in morphological and biochemical changes and finally in functional disturbances [for review, see 4]. In this context it is interesting to quote out of the studies of the United Kingdom [2] the corresponding chapter on vitamin C:

'Many studies have raised the question whether vitamin C has beneficial effects on normal human subjects at intakes and tissue levels considerably greater than those needed to prevent or cure scurvy. These studies have included examination of indices as diverse as histamine removal; cholesterol turnover; physical working capacity; immune function; male fertility; gingival collagen; nitrosamine and carcinogenesis prevention and selenium or iron utilization. Despite scientific concern about such questions, it is impossible to base estimates of requirements directly upon the evidence of these studies, partly because the evidence is conflicting in many areas, partly because those studies which have noted a positive benefit of vitamin C supplements have not defined the minimum dietary requirement needed to achieve it, and partly because specific design features have made the interpretation difficult in many cases. Because of these and other difficulties, the Panel decided to base their estimates of vitamin C requirements mainly upon the prevention of scurvy, on vitamin C turnover studies, and on biochemical indices of vitamin C status in man.'

We have a very similar situation for many of the vitamins of the B group that are all transformed to coenzymes. Because these coenzymes act at multiple sites of our metabolism, it is to be expected that many functions become reduced when food intake is unbalanced. For example, physical working capacity has been attributed to the marginal lack of vitamins C, B_2, B_6, changes in mental behavior to vitamins B_1, B_2, C and folic acid and lack of immune response to marginal lack of vitamin A, B_6 and C [4]. From the multiple action of each of these vitamins as coenzymes it is obvious that several of them are responsible for such a functional change and it has therefore been difficult to relate any of these functional changes to any single vitamin. Moreover, some of these functional changes may also be related in part to other disturbances of the body independent of the vitamin intake. From a scientific point of view it is therefore easy to understand that the decreasing functions as a result of marginal deficiencies are difficult to be included in a RDA value of a single

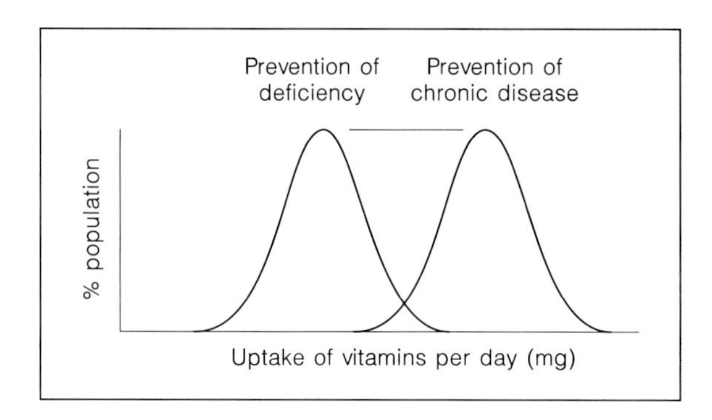

Fig. 2. Vitamin uptake and the prevention of deficiency and chronic disease.

vitamin. On the other hand, we know that an increased uptake of vitamins at levels somewhat higher than the RDAs may help to overcome such functional changes. Some important examples will be discussed during this workshop.

In recent years, evidence has accumulated showing that some vitamins as well as β-carotene may play an important prevention role in the reduction of risk of certain diseases such as cancer, cardiovascular diseases, cataracts and neural tube defects. We are dealing here actually with a completely new potential function of some of the vitamins. A lot of this evidence will be summarized in this symposium and as in the case with the marginal vitamin deficiencies, it will become more and more evident that 'traditional' vitamin intakes as recommended by the RDAs may not be sufficient for the specific preventive actions of vitamins and β-carotene (fig. 2).

In view of the fact that practically no cases of classical vitamin deficiencies are occurring in Western Europe today, the question can be raised if the present RDA concept is still a useful one. Therefore, one of the main topics of this symposium will be to discuss the issue whether the present RDA concept for the determination of RDAs should be changed or whether eventually new independent recommendations for the prevention of some of the well-described marginal deficiencies and especially for the reduction of the risk of chronic disease should be worked out.

The discussion should take into account that the general public more and more knows about the new developments in the prevention of neural tube defects as well as of chronic diseases. We as scientists are asked to give our recommendations in consideration of what we know today. This is our public duty.

To discuss all these questions, not only very well known scientists but also members from official national European and American organizations being concerned and also responsible for new recommendations are participating. We therefore hope that also some consequences with regard to public health and public statements can be discussed at this workshop as well.

Finally I would like to mention that this workshop is the first one of the recently established European Academy of Nutritional Sciences. This new organization has succeeded the former Group of European Nutritionists (GEN).

References

1 Committee on the Tenth Edition of the RDAs, Food and Nutrition, National Research Council: Recommended Dietary Allowances, ed 10. Washington, National Academy Press, 1989, p 10.
2 Department of Health: Dietary Reference Values for Food Energy and Nutrients for the United Kingdom, 1991. London, HMSO, 1992.
3 Commission of the European Communities: Food – Science and Techniques. Directorate-General Internal Market and Industrial Affairs, 1992.
4 Brubacher GB: Scientific basis for the estimation of the daily requirements for vitamins; in Walter P, Stähelin H, Brubacher G (eds): Elevated Dosages of Vitamins. Bern, Huber, 1989, pp 3–11.

Prof. Paul Walter, Swiss Vitamin Institute and Department of Biochemistry, University of Basel, Vesalgasse 1, CH–4051 Basel (Switzerland)

Walter P (ed): The Scientific Basis for Vitamin Intake in Human Nutrition.
Bibl Nutr Dieta. Basel, Karger, 1995, No 52, pp 7–19

History and Classical Functions of Vitamins

Klaus Pietrzik, Jutta Dierkes

Department of Pathophysiology of Human Nutrition,
Institute of Nutritional Science, Bonn, Germany

Vitamin deficiency diseases such as scurvy, night blindness or beriberi have been known since ancient times. Around 3,500 years ago, it was empirically known that night blindness or scurvy could be treated and even healed by certain foods. The modern era of vitamin research started at the end of the 19th century around 100 years ago with the induction of deficiency diseases such as beriberi in animals, leading to the concept that small amounts of accessory growth factors are necessary for growth and life.

Early in the 20th century, vitamin A and thiamin were the first to be discovered in animal trials. From these investigations, the term 'vitamin' was created by Casimir Funk in 1912, using the two verbs *vita* which comes from Latin and means life and *amine* because of the assumed chemical structure. It was believed that all these growth factors are of similar chemical structure. Though amino groups are not constituents of vitamins in any case, the term vitamin was accepted for all those growth factors in our food which are organic compounds and essentially needed for life in small amounts.

During the next decades, the other vitamins were discovered, their chemical structures elucidated, the substances were isolated and synthesized, ending in 1972 with the synthesis of vitamin B_{12}. Several Nobel prizes have been given to scientists for vitamin research (table 1).

In the early decades of vitamin research there was a lot of confusion concerning the essential character of those nutrients. A general accepted nomenclature of vitamins was not available (table 2). In the meantime there was agreement that only 13 vitamins have to be regarded as essential nutrients (see below).

Table 1. Nobel prizes for vitamin research

Prizewinner	Year	Topic
A.D.R. Windaus	1928 chemistry	sterins and vitamin D
C. Eijkman	1929 medicine	thiamin
F.G. Hopkins	1929 medicine	thiamin
P. Karrer	1937 chemistry	carotinoids and flavines
W.N. Haworth	1937 chemistry	carbohydrates and vitamin C
A. Szent-Györgyi	1937 medicine	vitamin C
R. Kuhn	1938 chemistry	vitamins and carotinoids
H.C.P. Dam	1943 medicine	vitamin K
E.A. Doisy	1943 medicine	vitamin K
F.A. Lipmann	1953	coenzyme A and pantothenic acid
H. Krebs	1953	coenzyme A and pantothenic acid
D. Hodgkin	1964 chemistry	vitamin B_{12}

The more the essential character of these nutrients became evident, the more there was a need to solve the question how much of these substances is necessary to maintain health.

The first recommendations for dietary intake were elaborated, founded on the basic knowledge on biochemical functions which were now available. However, as early as in 1862 the first recommendations at all were proposed to prevent starvation disease during the cotton famine in Lancashire. 'What is the least cost per head per week for which food can be bought in such quantity and in such quality as will avert starvation-disease from the unemployed population?' [1].

In 1933, the first British Medical Association's standards were formulated to maintain health and working capacity during depression. The League of Nations standards (1935–1938) were designed 'to marry health and agriculture'.

The Food and Nutrition Board was established in the United States in 1940. Five years later the Food and Nutrition Board published the first Recommended Dietary Allowances (RDAs) for six of the vitamins. The original purpose of the recommendations was to prevent disease or to maintain health. The RDAs have been revised about every 5 years and the 10th edition was published in 1989 [2]. Dietary allowances are now published by an increasing number of countries. It was a good idea to choose the term RDA instead of standard so any implication of finality could be avoided. Though there was

Table 2. The confusion in vitamin nomenclature

Vitamin A	
Vitamin A_2	Old name for dehydroretinol
Vitamin B_c	See folic acid
Vitamin B_p	Called the antiperosis factor for chicks, but can be replaced by manganese and choline
Vitamin B_T	An essential dietary factor for the mealworm, *Tenebrio molitor*
Vitamin B_w	Or factor W; probably identical with biotin
Vitamin B_x	Nonexistent; has been used in the past for both pantothenic acid and p-aminobenzoic acid
Vitamin B_1	Thiamin
Vitamin B_2	Riboflavin
Vitamin B_3	Nonexistent; term once used for pantothenic acid and sometimes, quite wrongly, used for niacin
Vitamin B_4	Name given to what was later identified as a mixture of arginine, glycine and cystine
Vitamin B_5	Name given to a substance later presumed to be identical with vitamin B_6 or possibly nicotinic acid
Vitamin B_6	Generic descriptor for three derivatives of 2-methylpyridine: pyridoxine, pyridoxal, pyridoxamine
Vitamin B_7	When a new factor was discovered which was claimed to be essential for chick growth and feathering, the claimant stated that as nine factors were known the new factors should be called vitamins B_{10} and B_{11}. In fact, the B vitamins had been numbered only up to B_6, hence B_7, B_8 and B_9 have never existed
Vitamin B_8	See vitamin B_7
Vitamin B_9	See vitamin B_7
Vitamin B_{10}	The names B_{10} and B_{11} were given to two factors claimed to be essential for chick growth and feathering; they were later shown to be a mixture of vitamin B_1 and folic acid
Vitamin B_{11}	See vitamin B_{10}
Vitamin B_{12}	Cobalamin
Vitamin B_{13}	See orotic acid; not an established vitamin
Vitamin B_{14}	Not an established vitamin; a substance found in human urine which increases the rate of cell proliferation in bone marrow culture
Vitamin B_{15}	Pangamic acid
Vitamin B_{16}	This term has never been used
Vitamin C	Ascorbic acid
Vitamin D	Calciferol
Vitamin E	Tocopherol
Vitamin F	Essential fatty acids
Vitamin G	Obsolete name for vitamin B_2
Vitamin H	See biotin
Vitamin K	
Vitamin L	Vitamin L_1 and L_2 are factors in yeast said to be essential for lactation; they have not become established
Vitamin M	See folic acid
Vitamin P	Name formerly given to a group of plant flavonoid substances which affect the strength of the walls of the blood capillaries. Now considered that the effect is pharmacological and that they are not dietary essentials; sometimes called 'bioflavonoid'
Vitamin PP	See nicotinic acid
Vitamin T	Factor found in insect cuticle. Also known as torulitine. Said to be a mixture of folic acid, vitamin B_{12} and desoxyribosides and not a new factor
Pantothenic acid	
Folic acid	
Niacin	
Biotin	

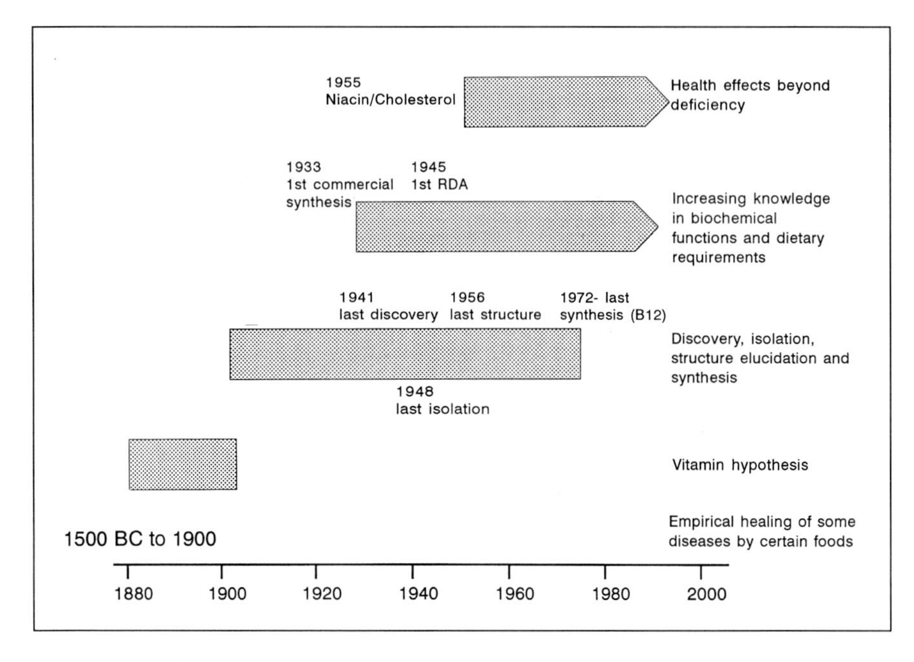

Fig. 1. Periods in the history of vitamins [modified from 3].

increasing knowledge on the function of vitamins during the last 50 years, only poor changes happened concerning the recommended levels of intake. However, newly discovered functions of vitamins which are not related to their classical role as coenzymes, are not included in the RDAs. The old view that vitamins are regarded to prevent vitamin deficiency disease is still the basis for most of the national RDAs.

During the last decades, other functions of vitamins have been discovered. In 1992, Machlin [3] stated that this period in vitamin research started in 1955, when the cholesterol-lowering effect of high-dose niacin was observed for the first time. Since then, vitamin research is rapidly growing and today it is known that vitamins have significant health effects beyond preventing deficiency diseases. There is increasing knowledge that vitamins have many important non-coenzymatic functions such as antioxidant activity, hormone-like and regulatory roles [3].

In view of these new developments it is necessary to re-examine the criteria for the establishment of RDAs. But before the new views are discussed, it might be helpful to look at the old ones. Some events in the history of vitamins as well as classical functions of vitamins will be highlighted in this review (fig. 1, table 3).

Table 3. Classical and potential roles of vitamins

Vitamin	Classical role	Potential role
Vitamin C	Hydroxylation reaction	In vivo antioxidant
β-Carotene	Provitamin A	Antioxidant
		Immune function
Vitamin K	Clotting factor	Calcium metabolism
Vitamin D	Calcium absorption	Differentiation and growth
	Mineralization of bone	Immune function
Vitamin B_6	Coenzyme	Steroid regulation
		Immune function
Niacin	Coenzyme	Lipid-lowering
Folate	One-carbon metabolism	Homocysteine metabolism
		Prevention of neural tube defects

Vitamin A

Night blindness was probably the first recognized nutritional disease which was already known in ancient Egypt. They recommended the topical application of juice squeezed from cooked liver to cure night blindness. Later in ancient Greece, the oral ingestion of liver and the topical application of this juice was recommended.

In the early years of our century, it could be shown that butter or egg yolk, but not lard, contained a lipid-soluble factor necessary for the growth of rats, and Stepp in Germany discovered one of these 'minimal qualitative factors' as a lipid. Vitamin A itself was discovered in 1909 and isolated in 1931. McCollum introduced the term 'fat-soluble A' in 1916. Classical functions of this vitamin were attributed to growth and tissue differentiation. The role of vitamin A in the visual cycle was basically clarified by George Wald in 1935, when he demonstrated the involvement of vitamin A in the photopigment formation in the retina of the eye [4, 5].

Vitamin D

Vitamin D deficiency disease, rickets, was already described in 1645. From this description it should take more than 350 years until the biochemical background was elucidated. Before vitamin D was discovered, rickets was thought to be caused by heredity or syphilis. The lack of sunlight and the lack of a nutritional factor have been associated with the occurrence of rickets during the second decade of our century.

The close relationship between ultraviolet light and a nutritional factor was recognized by experiments of Steenbock and Black in 1924. They found that irradiated food which was fed to rachitic rats was effective to cure rickets, but food which was not irradiated was without an effect. One year later it could be demonstrated that also irradiated skin contains a factor capable to prevent rickets. These observations led to the isolation and the elucidation of the chemical structure of vitamin D.

Classical functions of this vitamin were related to calcium absorption from the gut and calcium homeostasis interacting with parathyroid hormone and calcitonin. Vitamin D was believed to be the active agent, until it became evident that a metabolite (24, 25-$(OH)_2-D_3$) was 13–15 times as effective as vitamin D in stimulating intestinal calcium absorption and 5–6 times as effective in elevating serum calcium levels. Thus, this new metabolite appeared to be the biologically active form of vitamin D [6]. In modern nutrition, vitamin D is no longer regarded as a classical vitamin but as a prohormone, but traditionally it is classified as a vitamin.

Vitamin E

Vitamin E was discovered in the early 1920s when Herbert Evans and Katherine Bishop started a series of investigations on the influence of nutrition on reproduction in the rat. They discovered that rats failed to reproduce when fed a rancid lard diet unless lettuce or whole wheat were added to the diet.

The letter E was chosen to designate the factor following the earlier recognized vitamin D. After purification the factor was given the name tocopherol from the Greek *tocos*, meaning childbirth, and the verb *pherein*, to bring forth. Vitamin E is now accepted as the generic term for a group of tocol and tocotrienol derivates possessing some degree of vitamin activity. The most active compound is α-tocopherol.

Classical deficiency symptoms in chicks and guinea pigs such as encephalomalacia and muscle dystrophy were described in the third decade of our century. In 1931, Olcott and Mattill were the first to discover the antioxidant function of vitamin E. However, this vitamin has been something of a dilemma to nutritionists. Deficiency symptoms in humans similar to those in animals have not been established. It was only in 1968 that vitamin E was established as an essential factor in the human diet by the Food and Nutrition Board. The most accepted explanation for its activity is that it acts as a physiological antioxidant stabilizing various membranes and tissue components. In this role the vitamin aids the body in maintaining its normal defenses against disease and environmental insults. In absence of vitamin E, oxidative damage occurs to many cells, particularly the red blood cells and muscle and nerve cells [7].

Vitamin K

The discovery of vitamin K in 1929 was the result of a series of experiments by Henrik Dam [8]. He noted that chicks ingesting diets that had been ether-extracted to remove the sterols developed subdural or muscular hemorrhages and that blood taken from these animals clotted slowly. In 1935 he proposed that this antihemorrhagic factor was a new fat-soluble vitamin that he called vitamin K from the German word *Koagulation*. Vitamin K was isolated from plants by Dam and Karrer, and in 1939 it was synthesized by Doisy, Karrer and Fieser (9).

The regulation of blood clotting was elucidated only in the second half of our century, and it could be shown that also the synthesis of the blood clotting factors VIII, IX and X is dependent on vitamin K. In 1974, τ-carboxyglutamic acid, a so far unknown amino acid, was detected, and its synthesis was also found to be vitamin K-dependent [10].

Vitamin C

Scurvy, the vitamin C deficiency disease, was already described in papyrus Ebers in 1550 BC, and was known in ancient Greece and Rome. It was endemic in Northern Europe during winters when fresh fruits and vegetables were unavailable. Scurvy influenced the course of history, spontaneously ending many military campaigns and long ocean voyages by fatal outbreaks when rations became depleted of vitamin C.

During the times, treatment against scurvy was as often described as forgotten [11]. Newfoundland Indians consumed evergreen needle extracts in winter and various bulbs, early shoots, and leaves in spring to avoid scurvy. In 1570, Captain James Lancaster prevented scurvy by giving crew members two jiggers of lemon juice daily. In 1753–1757, Dr. James Lind published his 'Treatise of the Scurvy', which showed the disease to be a dietary deficiency from lack of fresh vegetables. During his voyages around the world (1772–1775), Captain Cook, following Lind's report, maintained a healthy crew with fresh vegetables and lemon juice in the diet. Buchan's Domestic Medicine 12th edition from 1791 described scurvy as follows:

'This disease may be known by unusual weariness, heaviness, and difficulty of breathing, especially after motion; rotteness of the gums, which are apt to bleed on the slightest touch; a stinking breath; frequent bleeding at the nose; crackling of the joints; difficulty of walking; sometimes a swelling and sometimes a falling away of the legs, on which there are livid, yellow, or violet-coloured spots; the face is generally of a pale or leaden colour. As the disease advances, other symptoms come on; as rotteness of the teeth, haemorrhages, or discharges of blood from different parts of the body, foul obstinate ulcers, pains in various parts, especially about the breast, dry scaly eruptions all over the body, etc. At last a wafting fever comes on, and the miserable patient is often carried off by a dysentery, a diarrhoea, a dropsy, a palsy, fainting fits, or a mortification of some of the bowels.'

Hirsch (1860) assumed that scurvy could be a deficiency of a nutritional factor. Vitamin C was then discovered during feeding trials with guinea pigs, and it was finally isolated in 1926, when Szent-Györgyi, who was working on oxidation processes in the adrenal cortex, isolated crystalline vitamin C. Classical functions of vitamin C relate to hydroxylation processes and insofar it is involved in the biosynthesis of collagen and carnitine as well as in the catecholamine synthesis and tyrosine degradation. Apart from these classical functions, the antioxidative potency of vitamin C additionally will be understood as a classical function, but the health-preventive potential of antioxidant function of vitamin C was recently understood and this has to be discussed in context with RDAs [12].

Thiamin

The first precise description of beriberi, the typical thiamin deficiency disease, was given by a Dutch physician in Java in 1630. Beriberi was a common disease particularly in Asia. It was Christian Eijkman who discovered that a polyneuritis resembling beriberi could be produced in chickens fed on polished rice. He further demonstrated (1900) that this polyneuritis could be prevented or cured by feeding rice bran. This was really the first experimental characterization of a nutritional deficiency [13].

When Casimir Funk obtained a crystalline substance from rice bran extracts (1911), he was convinced that he had isolated the active antiberiberi principle. He coined the term vitamin since this factor appeared to possess an amine function. He thus coined the name 'vitamine', or an amine essential for life.

The chemical structure of thiamin was published in 1935 by Williams, and the synthesis of this vitamin was shown in the same year. Physiological functions of thiamin were described by Thompson and Johnson in 1935. They demonstrated a high blood pyruvate level in B_1 deficiency. The only known biologically active form of thiamin is its diphosphate ester (thiamin pyrophosphate; TPP). TPP acts as coenzyme in reactions in the pyruvate metabolism, resulting in coenzyme A (CoA) formation.

In the transketolase reaction the TPP-dependent enzyme reacts with ketosugars and breaks the carbon-carbon bond between C-2 and C-3 to form thiamin diphosphate glycolaldehyde which is transferred into the pentose or hexose phosphate shunt (pathway for oxidation of glucose).

Thiamin is also involved in nerval functions. Nerves contain a rather constant and significant level of thiamin (triphosphate $\sim$10%) and stimulation of nerves results in a decrease in the level of thiamin. Though there is no doubt on the neurological involvement of thiamin, the mode of action is not definitely understood.

Riboflavine

After the discovery of vitamin B_1, a second nutritional factor was searched which was assumed to be a heat-stable factor necessary to prevent pellagra. Then vitamin B_2 was discovered in 1920 and isolated in 1933 from yeast, egg and whey. The chemical structure was elucidated in 1935. Different biological forms were described under which the most important ones are riboflavine-5'-phosphate (FMN) and flavine adenine dinucleotide (FAD). Both these active forms (FMN and FAD) form the prosthetic groups of a large number of enzymes which act as hydrogen transfer agents particularly in the metabolism of fatty acids and amino acids [14].

Niacin

Pellagra, the deficiency disease of niacin, was first described during the 18th century, following the introduction of maize in Europe. However, people living in Central America did not develop pellagra although maize was a staple food in their diet. In 1920, the relationship between maize intake and pellagra was elucidated by Goldberger. But the discovery that niacin can prevent pellagra was only in 1937. It became evident that pellagra is a combined deficiency of niacin and the amino acid tryptophan.

Nicotinic acid was already known since 1867. Its derivative nicotinamide has similar activity in humans. It can be synthesized in the mammalian body from the tryptophan and belongs to the B-vitamin complex. Niacin, the name that includes both substances with vitamin activity, is primarily involved in reactions that generate energy in tissues by the biochemical conversion of carbohydrates, fats and proteins. The active compounds are two coenzymes, nicotinamide adenine dinucleotide (NAD) and nicotinamide adenine dinucleotide phosphate (NADP).

Biotin

E. Wilders discovered in 1901 that yeast requires a special growth factor which he named 'bios'. Later it became evident that bios was the mixture of three compounds, from which bios IIB was biotin. Biotin is a water-soluble member of the B-vitamin complex. Deficiency of biotin causes lesions of the skin, and the vitamin is also known as vitamin H, derived from the German word *Haut*. It occurs in eight different forms, but only one, D-biotin, has full vitamin activity.

Biotin plays a key role in the metabolism of carbohydrates, fatty acids and amino acid, acting as coenzyme in carboxylation reactions.

Pantothenic Acid

'Bios', discovered by E. Wilders in 1901, was the mixture of three biological active compounds. One of them, bios IIA, was 30 years later identified to be pantothenic acid. The name refers to the Greek words meaning 'from everywhere'. Pantothenic acid is part of CoA, and is involved in reactions that supply energy, in the synthesis of cholesterol, hormones, neurotransmitters, phospholipids and porphyrin. Furthermore, it participates in acyl carrier protein, and insofar it is involved in the synthesis of fatty acids.

Pyridoxine

During the first years of our century, the differentiation between the many effects of the 'B-complex' was complicated. After the discovery of vitamin B_1 and vitamin B_2, a further nutritional factor was assumed to exist. György could demonstrate that this factor could prevent dermatitis acrodynia in rats. This was the discovery of vitamin B_6 in 1934. Four years later, vitamin B_6 was isolated in different laboratories. Two other natural forms of vitamin B_6, pyridoxal and pyridoxamine, were shown to exist in 1945. The biologically active form of vitamin B_6 is pyridoxal-5'-phosphate. The classical function is that of a coenzyme of several enzymes which play a vital role in amino acid metabolism. Apart from that, vitamin B_6 is involved in the synthesis of biogenic amines for brain activity.

New functions of this vitamin have been described with regard to immunological processes [14] and steroid-receptor complexes [15].

Vitamin B_{12}

In 1849, pernicious anemia was discovered as a disease of its own by Addison, an English physician. No treatment of this fatal disease was known until 1926 when Minot and Murphy [16] demonstrated that the disease could be cured by the intake of liver. In 1929, Castle showed that the intestinal absorption of the antipernicious anemia principle in liver (extrinsic factor) required prior binding to an 'intrinsic factor' secreted by the stomach. It took around 20 years until the active component present in liver concentrates, which was then called the 'antipernicious anemia factor', was isolated (1948). The synthesis of vitamin B_{12} required a number of years and was successful only in 1972.

In the human body, two enzymes are known to be dependent on vitamin B_{12}. These are the homocysteine-methyltransferase and the methymalonyl-CoA-racemase [17]. Due to the homocysteine-methyltransferase reaction, there is a close correlation to folate, as folate is involved in the same biochemical pathway.

Folic Acid

In 1930, Lucy Wills reported 'pernicious anemia of pregnancy' which she observed in pregnant women in Bombay and which could be cured by yeast extracts, in contrast to Addisonian pernicious anemia. Monkeys could be made anemic by feeding them with a deficient diet. Chicks were not growing if fed a semisynthetic diet although this diet was enriched with all the known vitamins [4]. These observations led to the discovery and isolation of folic acid. The molecule was isolated from spinach, which led to the name folic acid (latin *folium* = leaf). Folic acid was synthesized in 1946.

Bacterial studies revealed the essential character of folic acid, since purines or thymine could only partly replace folic acid or one of its compounds, *p*-aminobenzoic acid (PABA). The outstanding role of folic acid in one-carbon metabolism was elucidated in studies with radiolabeled formate and formaldehyde.

New Views on Vitamins

Some of these classical functions of vitamins (coenzymatic function of vitamin B_{12} and vitamin B_6; transfer of one-carbon units by folate) can be demonstrated in the metabolism of homocysteine. In general, homocysteine is transformed to cystathionine, respectively methionine; in both cases, vitamin-dependent enzymes are needed. The vitamin B_6-dependent cystathionine-β-synthase forms cystathionine, which is needed for cysteine and cystine synthesis. On the other hand, in the regeneration of methionine from homocysteine, vitamin B_{12} is needed as a coenzyme in the homocysteine-methyltransferase enzyme system. Apart from these coenzymatic functions of vitamin B_{12} and vitamin B_6 in homocysteine metabolism, another vitamin is involved which is folate as transfer system for one-carbon units.

These functions are regarded to be classical ones, but very recently it became apparent that homocysteine itself is of great importance with regard to atherosclerosis and neural tube defects [18, 19]. An increase of homocysteine is regarded to be harmful in respect of these diseases. On the other hand, lower homocysteine levels could be preventive. But these circumstances recently understood by scientists are not regarded to be a basis for RDAs. If we compare the older RDAs with the newer ones, no significant changes have happened.

Though the aim was to guarantee perfect health, the basis for RDAs is the historical perspective (the coenzymatic functions and prevention of deficiency), but in the meantime there is a lot of scientific evidence for newer roles of vitamins (antioxidant function, hormone-like action, regulatory role). If the guarantee of perfect health is the appropriate definition of RDAs, it can be concluded that the requirements are clearly higher than those for the prevention of clinical disorders.

In animals it is known that levels of vitamins in excess of those required purely for the maintenance of health lead to improved performance. The evidence for this in humans has been regarded to be satisfactory in the past. In future there is an increasing need to include our actual knowledge of additional functions in recommending vitamin intakes.

References

1 Truswell AS, et al: Recommended dietary intakes around the world. Nutr Abstr Rev Clin Nutr 1983;53:A939.

2 National Research Council (ed): Recommended Dietary Allowances. Washington, National Academy Press, 1989.

3 Machlin LJ: Beyond deficiency. Introduction. Ann NY Acad Sci 1992;669:1–6.

4 Friedrich W (ed): Handbuch der Vitamine. München, Urban & Schwarzenberg, 1987.
5 Machlin LJ (ed): Handbook of Vitamins. New York, Dekker, 1991.
6 Norman AW (ed): Vitamin D, The Calcium Homeostatic Steroid Hormone. New York, Academic Press, 1979.
7 Chow CK: Vitamin E and blood. World Rev Nutr Diet. Basel, Karger, 1985, vol 45, pp 133–166.
8 Dam H: Cholesterinstoffwechsel in Hühnereiern und Hühnchen. Biochem Z 1929;215:475–492.
9 Almquist HJ: The early history of vitamin K. Am J Clin Nutr 1975;28:656–659.
10 Stenflo J, Fernlund P, Egan W, Roepstorff P: Vitamin K dependent modifications of glutamic acid residues in prothrombin. Proc Natl Acad Sci USA 1974;71:2730–2733.
11 Carpenter KJ (ed): The History of Scurvy and Vitamin C. Cambridge, Cambridge University Press, 1986.
12 Counsell JN, Hornig DH (ed): Vitamin C (Ascorbic Acid). London, Applied Science Publishers, 1981.
13 Gubler CJ: Thiamin; in Machlin LJ (ed): Handbook of Vitamins. New York, Dekker, 1991, vol 6, pp 233–237.
14 Rall LC, Meydani SN: Vitamin B_6 and immune competence. Nutr Rev 1993;51:217–225.
15 Allgood VE, Cidlowski JA: Novel role for vitamin B_6 in steroid hormone action: A link between nutrition and the endocrine system. J Nutr Biochem 1991;2:523–534.
16 Minot GR, Murphy WP: JAMA 1926;91:923.
17 Schneider Z, Stroinsky A (ed): Comprehensive Vitamin B_{12}. Berlin, de Gruyter, 1987.
18 Clarke R, Daly L, Robinson K, Naughten E, Cahalane S, Fowler B, Graham I: Hyperhomocysteinemia: An independent risk factor for vascular disease. N Engl J Med 1991;324:1149–1155.
19 Steegers-Theunisssen RPM, Boers GHJ, Trijbels FJM , Eskes TKAB: Neural tube defects and derangement of homocysteine metabolism. N Engl J Med 1991;324:199.

Dr. Klaus Pietrzik, Department of Pathophysiology of Human Nutrition,
Institute of Nutritional Science, Endenicher Allee 11–13, D–53115 Bonn (Germany)

Walter P (ed): The Scientific Basis for Vitamin Intake in Human Nutrition.
Bibl Nutr Dieta. Basel, Karger, 1995, No 52, pp 20–29

Iron and Vitamins[1]

Leif Hallberg

Department of Internal Medicine, Clinical Nutrition Section,
University of Göteborg, Sweden

Several nutrients are required for normal erythropoiesis and anemia may thus be a feature of several nutritional deficiencies. Several vitamins are needed for normal erythropoiesis such as vitamin B_{12}, folic acid, vitamin C, vitamin E and vitamin A. I have been asked to address the topic 'iron and vitamins'. A direct relationship between iron and a vitamin has only been established for iron and vitamin C. The relationship between iron and vitamin A seems to be more indirect and complicated. There is also new interest in the role of vitamin A as a contributory factor in nutritional anemia in some populations. I will begin my discussion by briefly reviewing vitamin A and iron but devote most of my time to the relationship between iron and vitamin C, the latter being of considerable physiological and nutritional importance.

Vitamin A and Iron

It has been known for more than 70 years that vitamin A deficiency in rats induces changes in the bone marrow and is associated with anemia [1, 2]. Reports from studies in animals are conflicting, however, probably mainly related to the varying manifestations of the deficiency at different degrees of severity. In mild and early vitamin A deficiency, anemia is regularly observed. In more severe vitamin A deficiency combined with marked growth retardation, a microcytic, hypochromic polycythemia is seen which has been ascribed to dehydration. In some studies, but not all, increased contents of

[1] Supported by Swedish Medical Research Council project B94-04721-19A; The Swedish Council for Forestry and Agriculture Research, 50.0267/93 and 997/88L 113:3

iron in spleen, liver and bone marrow are reported. Plasma iron and transferrin saturation are not consistently changed. No information is available about red cell production, hemolysis or red cell survival. Several studies have found that iron absorption is increased. Thus, the experimental studies in animals show that hematopoiesis is altered but that these changes are far from consistent [3–5].

Experimental vitamin A deficiency in man, induced in middle-aged men, showed that anemia is regularly seen. The hemoglobin concentration is proportional to the serum vitamin A levels. Iron treatment had a short temporary effect, whereas treatment with vitamin A fully normalized hemoglobin levels [6]. The pathogenesis of the anemia has still not been established [7, 8].

Nutritional anemia is mainly caused by a deficiency of iron. Several observations in children and pregnant women in different parts of the world suggest, however, that a deficiency in vitamin A could be an important contributory factor in the anemia.

There is no evidence that vitamin A deficiency has a direct relationship to the metabolism of iron [8]. Mild vitamin A deficiency seems to induce a mild microcytic, hypochromic anemia resembling the findings in iron deficiency anemia. The cause, however, does not seem to be related to a primary change in the metabolism of iron or to a lack of iron, but rather to some kind of ineffective erythropoiesis. This would explain both the increased absorption of iron and the change in the distribution of iron in the body to storage compartments [9–11].

Vitamin C and Iron

It was demonstrated more than 40 years ago that ascorbic acid increased the absorption of nonheme iron [12]. Using the extrinsic tag method, several groups showed that the effect of ascorbic acid was marked and consistent [13–29, 31, 33–36, 38, 39]. No effect was seen on the absorption of heme iron. Several investigators have shown that the effect of ascorbic acid on nonheme iron absorption is strongly dose-related [20, 33]. It was also shown that the effect of synthetic ascorbic acid and naturally occurring vitamin C was the same [33]. The properties of the meal significantly influenced the magnitude of the response. The more inhibitors present in a meal, the more marked was the relative effect on iron absorption [33]. Phytate and iron-binding phenolic compounds are two strong dietary inhibitors of nonheme iron absorption. Their inhibiting effect, however, can be strikingly counteracted by ascorbic acid [22, 23, 35, 36, 38, 39]. Dietary ascorbic acid is thus an important component in the normal dietary balance between factors that inhibit and enhance iron absorption. The bioavailability of iron in weaning foods is considerably improved by ascorbic acid and is today almost generally used in most infant cereals [23].

Prolonged warming of a meal, which substantially decreases the content of ascorbic acid, also reduces iron absorption. Restituting the ascorbic acid content in such a meal also restitutes absorption, showing that the effect of heating is related to the destruction of ascorbic acid [26]. *In summary*, the studies of the effect of ascorbic acid on iron absorption enumerated are examples from the wide range of reports by different investigators consistently showing that ascorbic acid increases the absorption of dietary nonheme iron and that this effect is clearly dose-dependent. The ascorbic acid content in the diet has a determining effect on the bioavailability of iron and thus on iron balance in the population.

There are three studies, however, where the effect of ascorbic acid has not been as marked or the same as that expected by the investigators [30, 32, 37]. These results may therefore seem to contradict the conclusion that ascorbic acid has a significant effect on iron nutrition. The three studies were performed carefully and by experienced scientists. The results cannot be disregarded but need to be thoroughly examined in this analysis of the nutritional importance of ascorbic acid.

The common denominator of these three studies is that the effect of ascorbic acid on iron absorption was based on an expected increase in serum ferritin. This in turn was considered as a reliable measure of an expected increase in iron stores. In two of the studies, serum ferritin was the only measurement of the effect of ascorbic acid on iron balance [30, 32]. In one study several other measurements were made [37]. Therefore it is important to first examine the conditions for using serum ferritin to measure an increase in iron stores.

Serum ferritin is an indirect measure of iron stores and has been carefully validated for this purpose. During a positive iron balance, for example that induced by giving extra ascorbic acid with the main meals to a subject with iron deficiency anemia, the extra iron expected to be absorbed will first be used to restitute hemoglobin and tissue iron. Iron stores cannot be expected to increase and an increase of serum ferritin cannot be expected until the individual optimal hemoglobin has been attained. Another fact to be considered is that, in women with iron deficiency anemia due to heavy menstrual blood losses, it may not be sufficient to increase the absorption of iron from the diet with even large amounts of ascorbic acid in order to treat the anemia and balance the heavy iron losses.

Let us now address our attention to the three studies:

In one study by Malone et al. [32], 100 mg ascorbic acid was given with three main meals for 8 weeks to 25 young women. Five of these had subnormal serum ferritin values prior to the study. In a control group of 23 women given placebo, 9 had subnormal serum ferritin at the commencement of the study. During the study, serum ferritin increased by 4.3 μg/l, from 27 to 31.3 μg/l, in the treatment group, but by only 0.067 μg/l in the control group. The difference did not reach statistical significance ($0.10 > p > 0.05$). From the distribution of serum ferritin in young women it can be estimated, using the t test for a difference, that the total sample should have comprised more women in order to reach statistical significance, or that the study should have continued for a longer time.

In another study by Hunt et al. [37], iron absorption was measured using the chemical balance technic with adequate, long balance periods. Iron deficiency was induced in 11 healthy, premenopausal women by phlebotomies and an iron-poor diet. Between the 109th and the 149th day of the study, when hemoglobin values were significantly reduced and serum ferritin values were below 8.5 μg/l in all subjects, 500 mg ascorbic acid or placebo tablets were given 3 times daily with major meals. Ascorbic acid induced a significant increase in total iron absorption, in hemoglobin and plasma iron concentration, but not in serum ferritin. It should be noted that the hemoglobin values had not reached their normal values at the end of this balance period and that, as mentioned, therefore, one cannot expect serum ferritin to increase. This interpretation was also made by the authors of this meticulous and well-designed study. The conclusion is thus that ascorbic acid significantly increased iron absorption during the 5.5-week period.

In the third study by Cook et al. [30], 1,000 mg ascorbic acid was taken daily with each of two main meals for 16 weeks in a sample of 8 women and 9 men. In 9 subjects, the administration of vitamin C continued for 24 months. By and large, there was no change in serum ferritin in the normal subjects but there was an increase in the iron-deficient subjects, especially in the only male subject who was iron-deficient due to previous frequent blood donations. A very important study within the study was also made. Iron absorption from a test meal with low bioavailability was measured when the meal was given without or with 1 g of ascorbic acid. This measurement of the effect of ascorbic acid was made both before the study and 16 weeks later, in other words after 4 months of daily, high-dose ascorbic acid supplementation. On both occasions, the effect of ascorbic acid was very marked, showing a more than 5-fold increase in iron absorption. There was thus no sign of any exhaustion or weakening of the response to ascorbic acid in these subjects in spite of the high doses given.

The findings that after giving these high doses of ascorbic acid, there was little or no effect on the individual serum ferritin values and that there was still a marked effect on the absorption of iron from the test meal, may seem contradictory and unaccountable. None of the several explanations put forward by the authors were considered by them to be quite satisfactory. The results of this study have to some extent called into question the long-term usefulness of ascorbic acid as a promoter of iron absorption.

Some recent studies made in our laboratory may provide an explanation of these seemingly contradictory results. Recently, and for the first time using the extrinsic tag method, we measured the total amount of iron absorbed from the whole diet over periods of 5–10 days by uniformly labelling to the same specific activity all nonheme iron present in all meals. Heme iron absorption was calculated in each subject from the known relationship between the absorption of heme iron and the reference dose containing a 3-mg dose of inorganic iron. These studies were made in young, healthy, highly motivated women, senior students of dietetics, with no known infection within the preceding month that might have influenced the serum ferritin level. All meals were served under supervision in the laboratory. In each subject, menstrual iron losses were also measured so that individual iron requirements could be calculated. Moreover, the usual hematological measurements were made and serum ferritin was determined as a measure of iron stores. In two groups we served a diet which we considered to be optimal with respect to iron absorption, containing plenty of meat, fish and ascorbic acid-rich fruits and vegetables and only small amounts of inhibitors of iron absorption. Of the 20 main meals served in 10 days, 3 were fish meals and 17 were meat meals (3 of these were chicken). The meat meals contained on average 120 g meat. Each main meal contained 40–50 mg ascorbic acid. Total energy intake was 2,200 kcal (=918 MJ). Total daily iron intake was 13.3 mg out of which 1.4 mg was heme iron.

As expected, we found a close relationship between serum ferritin and total amounts of iron absorbed (fig. 1). The unexpected finding was that, when serum ferritin reached a level of about 50μg/l, the total amounts of iron absorbed from the two diets with high bioavailability were so low that they corresponded to, or were even lower than, the basal iron requirements – in other words the amounts needed to cover basal iron losses. In those given the diet with a lower bioavailability, the critical absorption was found at a serum ferritin of 25 μg/l. These findings imply that the regulation of the absorption of iron at this higher end of the iron status scale is very efficient and that it is impossible, at serum ferritin values exceeding the values mentioned, to achieve a positive iron balance. It is thus impossible to absorb more iron than the amounts needed to cover the basal iron requirements. This also means

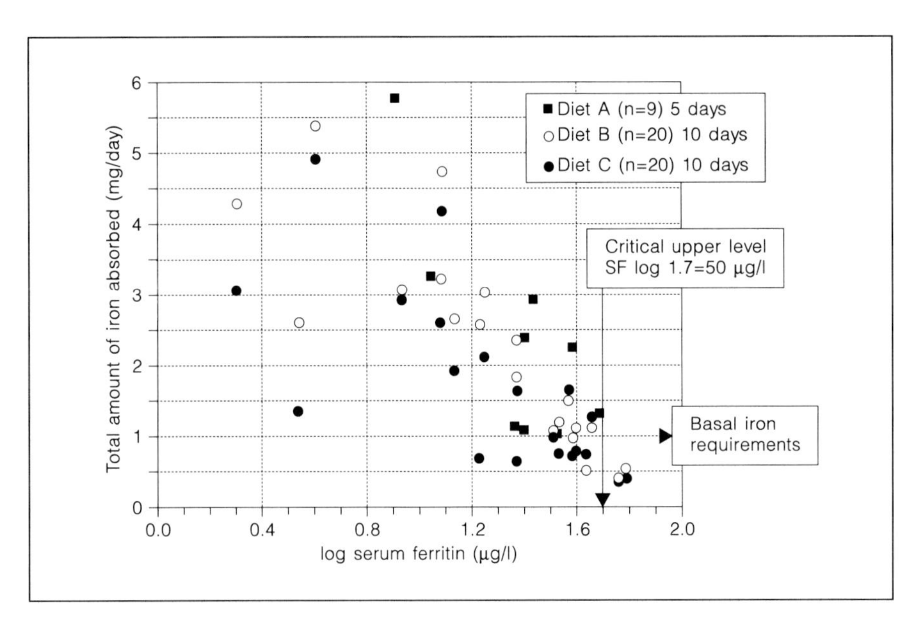

Fig. 1. Total amount of iron absorbed from whole diet in relation to serum ferritin. All nonheme iron in all meals was uniformly labelled with an extrinsic tracer to the same specific activity and heme iron was calculated from absorption of a reference dose of inorganic iron and the known relationship at different iron status to heme iron absorption. At serum ferritin values exceeding 50 µg/l, no subject absorbed more iron than the basal iron requirements. This limit thus constitutes the upper limit for any accumulation of iron in the body with these high bioavailability diets (diets A and B). Diet C had a lower bioavailability of iron due to high intake of milk with main meals.

that it would be improbable for iron stores to increase up to more than about 400–500 mg in women by any diet and by diet alone, including diets with a high ascorbic acid content. It is probable that, in men, the corresponding figure for maximal iron stores would be around 700–800 mg due to their larger body size. These data fit nicely with measurements of iron stores using repeated phlebotomies over a long period, when proper correction is made for the increased absorption of dietary iron during the phlebotomy period.

It is well known that there are subjects in most populations with higher serum ferritin values than the ones mentioned. It is thus necessary to take into account the following facts. It is probably not the serum ferritin per se that directly influences iron absorption but rather some unknown ‘signal’ related to iron stores. Moreover, several factors besides iron status influence the concentration of serum ferritin. A number of diseases such as liver diseases, inflammatory diseases and infections markedly increase the concentration of serum ferritin. It is a recognized fact that even mild infections may increase serum ferritin for several weeks. Another, usually disregarded, fact to be considered, which leads to higher serum ferritin levels, is the declining hemoglobin mass commencing already in middle-aged and even younger people, related to the change towards a successively more sedentary lifestyle and the accompanying shift of iron from red cell mass to stores.

A comment should also be given to the maintained enhancing effect of ascorbic acid on the absorption of iron from the test meals in the study by Cook et al. [30]. These measurements were made in the morning after an overnight fast. The bioavailability of iron in these test meals was very low and absorption increased from only about 0.1 mg to approximately 0.4–0.6 mg. The findings made in our study do not imply that iron absorption ceases completely at a certain level of stored iron – only that the absorption is low and does not exceed the amounts needed to cover the basal iron requirements. The amounts of iron absorbed from the test meals were well below this physiological limit.

An important conclusion drawn by Cook et al. [30] from their studies was that even massive doses of vitamin C may not lead to progressive iron accumulation in iron replete subjects, except in those few who are homozygotes for hereditary hemochromatosis. Their interpretation was that the regulatory mechanisms controlling body iron stores override any pronounced alteration in food iron availability.

Our recent observations that there is a very strong control of the absorption of dietary iron when iron stores reach certain levels are fully consistent with these interpretations. This means not only that ascorbic acid is the most marked enhancer of nonheme iron absorption, but also that it is safe.

Iron and Dietary Requirements for Vitamin C

There is good biochemical evidence that iron has to be in a ferrous state during some critical phase of the absorption process. Even if most food iron is present in a ferric state, the presence of ascorbic acid in the diet is not an absolute condition for the absorption of dietary iron. There are probably sufficient amounts of other reducing substances in the gastrointestinal lumen to ensure that some iron is absorbed. It is difficult, however, to meet the physiological requirements of dietary iron in most subjects without the presence of rather considerable amounts of ascorbic acid in most meals.

In our studies presented in which we measured iron absorption from all meals over two 10-day periods, the content of ascorbic acid as analyzed in each of the two main meals was 40–50 mg. In each woman we also measured the individual iron requirements from menstrual losses of iron and the basal losses from body weight. In 20% of the women their iron requirements were not covered by the absorption from this good diet containing not only 40–50 mg ascorbic acid in each meal but also ample amounts of meat.

When we just moved the milk intake from breakfast and evening meals to the main meals the bioavailability of the dietary iron was reduced. Less iron was absorbed and the iron requirements were not covered in 30%.

It should be emphasized that the lower the energy expenditure and energy intake the more meat and the more ascorbic acid are required in the diet to achieve a sufficient bioavailability of the dietary iron.

From the dose-response curves and from the present studies it is evident that each main meal should contain not less than 50 mg ascorbic acid. If the meal contains much inhibitors, for example, from cereals rich in fiber and phytate, the intake of ascorbic acid with such meals should be higher and not less than 75–100 mg.

It is important that we realize that ascorbic acid has a crucial, physiological role in iron absorption. This role is not less important than its role in preventing scurvy. It is my firm conviction that if a nutrient has several physiological functions in the body, the one requiring the highest amounts should be used as the basis for setting the recommendations for dietary intake.

Conclusions

To conclude: (1) Ascorbic acid has a key role in the absorption of dietary nonheme iron. (2) Natural and synthetic ascorbic acid have the same effect. (3) Ascorbic acid has a log/log dose-effect relationship in iron absorption. (4) The absorption is *not* reduced by prolonged intake of higher amounts of ascorbic acid. (5) There is a very efficient regulatory system for the absorption of iron preventing the development of dietary iron overload in normal subjects. (6) Each main meal should contain at least 50 mg ascorbic acid – more if a meal contains much phytate, and more if energy expenditure is low.

References

1 Findlay GM, MacKenzie RD: The bone marrow in deficiency diseases. J Pathol 1922;25:402–403.
2 Wolbach SB, Howe PR: Tissue changes following deprivation of fat-soluble A vitamin. J Exp Med 1925;42:753–781.
3 Amine EK, Hegsted DM, Hayes KC: Comparative hematology during deficiencies of iron and vitamin A in the rat. J Nutr 1970;100:1033–1040.
4 Mejia LA, Hodges RE, Rucker RB: Role of vitamin A in the absorption, retention and distribution of iron in the rat. J Nutr 1979;109:129–137.

5 Staab DB, Hodges RE, Metcalf WK, Smith JL: Relationship between vitamin A and iron in the liver. J Nutr 1984;114:840–844.
6 Hodges RE, Sauberlich HE, Canham JE, Wallace DL, Rucker RB, Mejia LA, Mohanram M: Hematopoietic studies in vitamin A deficiency. Am J Clin Nutr 1978;31:876–885.
7 Mejia LA, Hodges RE, Arroyave G, Viteri F, Torun B: Vitamin A deficiency and anemia in Central American children. Am J Clin Nutr 1977;30:1175–1184.
8 Mejia LA, Arroyave G: The effect of vitamin A fortification of sugar on iron metabolism in preschool children in Guatemala. Am J Clin Nutr 1982;36:87–93.
9 Sijtsma KW, Van den Berg GJ, Lemmens AG, West CE, Beynen AC: Iron status in rats fed on diets containing marginal amounts of vitamin A. Br J Nutr 1993;70:777–785.
10 Wolde-Gebriel Z, West CE, Gebru H, Tadesse AS, Fisseha T, Gabre P, Aboye C, Ayana G, Hautvast JGAJ: Interrelationship between vitamin A, iodine and iron status in schoolchildren in Shoa region, Central Ethiopia. Br J Nutr 1993;70:593–607.
11 Suharno D, West CE, Muhilal Karyadi D, Hautvast JGAJ: Supplementation with vitamin A and iron for nutritional anaemia in pregnant women in West Java, Indonesia. Lancet 1993;342: 1325–1328.
12 Moore CV, Dubach R: Observations on the absorption of iron from foods tagged with radioiron. Trans Assoc Am Physicians 1951;64:245–256.
13 Cook JD, Layrisse M, Matinez-Torres C, Walker R, Monsen E, Finch CA: Food iron absorption measured by an extrinsic tag. J Clin Invest 1972;51:805–815.
14 Sayers MH, Lynch SR, Jacops P, Charlton RW, Bothwell TH, Walker RB, Mayet F: The effect of ascorbic acid supplementation on the absorption of iron in maize, wheat and soya. Br J Haematol 1973;24:209–218.
15 Sayers MH, Lynch SR, Charlton RW, Bothwell TH, Walker RB, Mayet F: Iron absorption from rice meals cooked with fortified salt containing ferrous sulphate and ascorbic acid. Br J Nutr 1974; 31:367–375.
16 Sayers MH, Lynch SR, Charlton RW, Bothwell TH, Walker RB, Mayet F: The fortification of common salt with ascorbic acid and iron. Br J Haematol 1974;28:483–495.
17 Layrisse M, Martinez-Torres C, Gonzales M: Measurement of total daily iron absorption by the extrinsic tag method. Am J Clin Nutr 1974;27:152–162.
18 Björn-Rasmussen E, Hallberg L: Iron absorption from maize. Effect of ascorbic acid on iron absorption from maize supplemented with ferrous sulphate. Nutr Metabol 1974;16:94–100.
19 Disler PB, Lynch SR, Charlton RW, Bothwell TH: Studies on the fortification of cane sugar with iron and ascorbic acid. Br J Nutr 1975;34:141–152.
20 Cook JD, Monsen ER: Vitamin C, the common cold and iron absorption. Am J Clin Nutr 1977; 39:235–241.
21 Derman D, Sayers M, Lynch SR, Charlton RW, Bothwell TH: Iron absorption from a cereal-based meal containing cane sugar fortified with ascorbic acid. Br J Nutr 1977;38:261–269.
22 Rossander L, Hallberg L, Björn-Rasmusssen E: Absorption of iron from breakfast meals. Am J Clin Nutr 1979;32:2484–2489.
23 Derman DB, Bothwell TH, MacPhail AP, Torrance JD, Bezwoda WR, Charlton RW, Mayet FGH: Importance of ascorbic acid in the absorption of iron from infant foods. Scand J Haematol 1980; 25:193–201.
24 Hallberg L: Effect of vitamin C on the bioavailability of iron from food; in Counsell JN, Hornig DH (eds): Vitamin C (Ascorbic Acid). London, Applied Science Publishers, 1981, pp 49–61.
25 Hallberg L, Rossander L: Effect of different drinks on the absorption of non-heme iron from composite meals. Hum Nutr Appl Nutr 1982;36A:116–123.
26 Hallberg L, Rossander L, Persson H, Svahn E: Deleterious effects of prolonged warming of meals on ascorbic acid content and iron absorption. Am J Clin Nutr 1982;36:846–850.
27 Gillooly M, Bothwell TH, Torrance JD, et al: The effects of organic acids, phytates and polyphenols on the absorption of iron from vegetables. Br J Nutr 1983;49:331–342.
28 Hallberg L, Rossander L: Improvement of iron nutrition in developing countries: Comparison of adding meat, soy protein, ascorbic acid, citric acid, and ferrous sulphate on iron absorption from a simple Latin-American type of meal. Am J Clin Nutr 1984;39:577–583.

29 Gillooly M, Torrance JD, Bothwell TH, MacPhail AP, Derman D, Mills W, Mayet F: The relative effect of ascorbic acid on iron absorption from soy-based infant formulas. Am J Clin Nutr 1984; 40:522–527.
30 Cook JD, Watson SS, Simpson KM, Lipschitz DA, Skikne BS: The effect of high ascorbic acid supplementation on body iron stores. Blood 1984;64:721–726.
31 Seshadri S, Bhade S: Haematological response of anaemic preschool children to ascorbic acid supplementation. Hum Nutr Appl Nutr 1985;39A:151–154.
32 Malone HE, Kevany JP, Scott JM, O'Broin SD, O'Connor G: Ascorbic acid supplementation: Its effect on body iron stores and white blood cells. Irish J Med Sci 1986;155:74–79.
33 Hallberg L, Brune M, Rossander L: Effect of ascorbic acid on iron absorption from different types of meals. Hum Nutr Appl Nutr 1986;40A:97–113.
34 Hallberg L, Brune M, Rossander-Hultén L: Is there a physiological role of vitamin C in iron absorption? Ann N Y Acad Sci 1987;498:324–332.
35 Hallberg L, Brune M, Rossander L: Iron absorption in man: Ascorbic acid and dose-dependent inhibition by phytate. Am J Clin Nutr 1989;49:140–144.
36 Tuntawiroon M, Sritongkul N, Rossander-Hultén L, Pleehachinda R, Suwanik R, Brune M, Hallberg L: Rice and iron absorption in man. Eur J Clin Nutr 1990;44:489–497.
37 Hunt JR, Mullen LM, Gallager SK, Nielsen FH: Ascorbic acid: Effect on ongoing iron absorption and status in iron-depleted young women. Am J Clin Nutr 1990;51:649–655.
38 Tuntawiroon M, Sritongkul N, Brune M, Rossander-Hultén L, Pleehachinda R, Suwanik R, Hallberg L: Dose-dependent inhibitory effect of phenolic compounds in foods on non-heme-iron absorption in men. Am J Clin Nutr 1991;53:554–557.
39 Siegenberg D, Baynes RD, Bothwell TH, Macfarlane BJ, Lamparelli RD, Car NG, MacPhail P, Schmidt U, Mayet F: Ascorbic acid prevents the dose-dependent inhibitory effects of polyphenols and phytates on non-heme-iron absorption. Am J Clin Nutr 1991;53:537–541.

Prof. Leif Hallberg, Division of Clinical Nutrition, Department of Medicine, University of Göteborg, Annedalsklinikerna, Sahlgren Hospital, S-413 45 Göteborg (Sweden)

Walter P (ed): The Scientific Basis for Vitamin Intake in Human Nutrition.
Bibl Nutr Dieta. Basel, Karger, 1995, No 52, pp 30–42

Nitrosamines and Vitamins

Peter I. Reed

Lady Sobell Gastrointestinal Unit, Wexham Park Hospital,
Slough, Berks., UK

N-nitroso compounds (NOC) are a major class of chemical carcinogens, several hundred of which have now been identified and found to be either mutagenic or carcinogenic in all 40 susceptible species tested, including mammals, reptiles, birds, amphibia and fish [1, 2]. These compounds show a remarkable degree of organ specificity varying with their chemical structure, species tested, dosage and route of administration. Organs in which cancers have been induced include the oesophagus, stomach, liver, lung and urinary bladder. The carcinogenicity, mutagenicity and teratogenicity of a large number of NOC have been extensively documented. Of 232 N-nitrosamines studied, 199 (86%) proved positive, while 91 out of 100 N-nitrosamides, compounds not requiring metabolic activation to exert their action, induced tumour formation [3]. Although the evidence is circumstantial, NOC almost certainly contribute to tumour incidence in man. Humans may be exposed to these carcinogens in several different ways including: (a) Formation in the environment and subsequent intake through food, water, air, dermal contact and industrial or consumer products. (b) Formation in the body from precursors either ingested in food or water or formed endogenously.

While it remains to be established whether environmentally derived NOC present a significant risk to humans, for at least one cancer such an association is most likely. There is strong epidemiological association in the southern USA of oral cancer with snuff dipping, as distinct from snuff sniffing or tobacco chewing [4]. The tobacco-specific N-nitrosamines, including N-nitrosonornicotine (NNN) and 4-(methylnitrosamino)-(3-pyridyl)-1-butanone (NNK) are by far the most potent and abundant carcinogens detected in oral snuff [5]. Nitrosation of nicotine during the curing, fermentation and ageing of tobacco leads to the formation of these and other tobacco-specific NOC [6], which experimentally have induced both benign and malignant tumours in the oronasal cavity, oesophagus and lungs of rodents [7].

NOC are synthesized in the human body and it has been suggested, backed by an increasing volume of supporting evidence, that the endogenous formation of NOC probably is an important factor in the aetiology of other cancers, including those of the stomach, oesophagus and bladder [8, 9]. A comparative study of dietary sources of potentially N-nitrosatable substrates [10] had demonstrated the enormous range of possibilities for in vivo NOC synthesis, mainly under acidic conditions. Man is exposed to various environmental nitrosating agents and nitrosatable compounds including amines, ureas, guanidines and carbamates derived from food, water or drugs. Nitrate, which can be reduced to nitrite (NO_2), is present in water, in most foods and notably in almost all, especially root vegetables, salted foods (such as those popular in Japan) and some cured meats. Roughly 25% of ingested nitrate appears in the saliva, where it can be reduced by many oral or gastric bacteria. Thus, NO_2 is a normal constituent of human saliva, its concentration depending largely on the nitrate intake. NO_2 can be formed also through bacterial reduction of nitrate in several sites in the body, including the mouth, stomach, oesophagus, bladder and colon under normal or pathological conditions or by heterotrophic nitrification from ammonia or other reduced nitrogen. The production of NO_2, an unstable and highly reactive ion, generally occurs under near neutral conditions, the peak activity occurring at pH 5–6.5. Therefore, it is probable that man is continuously exposed to the precursors that can give rise to the formation of NOC through nitrosation in vivo. While few data exist for evaluating human exposure to NOC in quantitative terms, it has been suggested that endogenous nitrosation is probably the largest single source of exposure to these compounds for the general population [11].

Since Druckrey et al. [12] suggested the possibility that carcinogenic NOC may be formed from precursor amino compounds and nitrate under the acidic conditions of the stomach, various chemical and biological procedures have been utilized to demonstrate that the in vivo nitrosation reaction does occur in animals. Sander and Bürkle [13] were the first to show that the reaction between the ingested secondary amines and nitrite could occur in vivo and could produce carcinogenic N-nitrosamines in laboratory animals. However, the ability for in vivo N-nitrosation to occur is determined by multiple factors such as the gastric juice pH, the quantities and nature of precursors involved or the presence of modifiers acting as catalysts or inhibitors (table 1). In addition, for a long time inadequate analytical techniques and lack of proper controls to ensure against the formation of artefacts during collection procedures and during analysis also resulted in misleading data. The development by Ohshima and Bartsch [14] of an effective and simple method for demonstrat-

Table 1. Catalysis and inhibition of N-nitrosation

Catalysis	Inhibition
Thiocyanate	Ascorbic acid
Bromide	α-Tocopherol
Chloride	Sulphite, sulphur dioxide
Formaldehyde	Azide
Pyridoxine (vitamin B_6)	Bromate
Phenols *(if nitrite in excess)*	Urea
Lecithin and other micelle-forming	Amidosulphonic acid
quaternary surfactants	Cysteine and other sulphydryl componds
Cell wall of microorganisms	Gallic acid
	Phenols *(if nitrite $<$ equimolar)*

ing the endogenous formation of N-nitrosoproline (NPRO), initially in animals and then in man, led to significant progress in this field. The method, based on the use of specified oral doses of nitrate (beetroot juice) and an amine (*L*-proline), and measuring the resulting NPRO in a 24-hour urine collection, is effective because NPRO is not metabolized, is not carcinogenic and can be measured quantitatively in the urine.

The NPRO test has been employed by many research workers since its publication in 1981; they have all confirmed that NOC can be formed endogenously in animals and man. These studies have demonstrated that urinary NPRO levels increase when nitrate and *L*-proline are taken orally, that the amount of NPRO formed depends on the nitrate intake and that the NPRO excretion returns to baseline levels when ascorbic acid is administered with a proline dose [14]. A molar ratio of ascorbic acid to nitrite (2:1) is sufficient to completely inhibit NPRO formation in vitro. However, ascorbic acid doses 20 times larger than the estimated gastric nitrite (assuming that 5% of the nitrate dose is reduced to nitrite) do not completely eliminate urinary NPRO [14]. That NPRO formation only takes place in an acidic stomach, at a pH <3.5, has been used as an argument against bacterially mediated N-nitrosation at an high pH being relevant in gastric carcinogenesis [15]. However, this viewpoint has been refuted by several authors [16–18]. Neither do dietary sources of preformed NPRO account for the excess urinary NPRO. It has been suggested that NPRO may be formed at some site(s) other than the stomach, ones that are probably inaccessible to ascorbic acid and perhaps via a mechanism other than that of acid catalysed nitrosation. Sites which have been proposed include the oral cavity, oesophagus, the infected urinary bladder [9] and colon. In all these organs, and under normal physiological conditions and at an high pH, bacterial growth can occur and thereby lead to the formation of NOC precursors and catalysis of the N-nitrosation reaction. It is still unknown, however, how many of these NOC could be formed in this manner and the degree to which in vivo N-nitrosation plays a role in human carcinogenesis.

Two other possible sources of endogenous nitrosating agents in addition to dietary nitrate have been suggested; they are atmospheric nitrogen oxides and nitric oxide (NO) produced endogenously by cells. NO is a labile species that can react rapidly with oxygen yielding NO_2, a known nitrosating and nitrating agent [19], which exists in equilibrium with the potent nitrosating agents N_2O_3 and NO_4. Subsequent reaction of these compounds to secondary amines would yield N-nitrosamines and the competing reaction with water would yield nitrite from nitrate.

The reaction of atmospheric nitrogen oxides, particularly nitrogen dioxide (NO_2), with endogenous amines, represents another possible nitrosation pathway. For instance, cigarettes are one such significant source of exposure; cigarette smoke contains as much as 1,000 ppm nitrogen oxides. Also, Mirvish et al. [20] were able to demonstrate in vivo formation of increased levels of N-nitrosomorpholine in rodents exposed by inhalation to nitrogen dioxide and gavaged with morpholine, while Garland et al. [21] reported a positive correlation between atmospheric NO_2 levels and N-nitrosodimethylamine (NDMA) excretion in humans. However, it could not be established whether this increase was due to additional nitrosation of dimethylamine in vivo or nitrosation in the air and subsequent inhalation of NDMA. However, there was no correlation between urinary NPRO and atmospheric NO_2 concentrations.

Bacteria have also been shown to be capable of mediating N-nitrosation and their role has been extensively considered in hypotheses regarding the aetiology of several cancers, in particular gastric cancer. In the now classical model of gastric carcinogenesis, Correa et al. [22] postulated that in the presence of gastric hypochlorhydria, especially when the pH is >5, increased levels of endogenous N-nitrosation occur because of gastric bacterial overgrowth, with NO_2 concentrations increasing correspondingly. Such overgrowth can also occur at other sites including the urinary bladder and vagina. It was at first thought that bacteria facilitate nitrosation primarily by reducing nitrate to nitrite, but subsequent studies have shown that bacteria act directly in amine nitrosation and that this activity is linked to nitrate reductase genes [23, 24].

However, there has been controversy concerning the relationship between intragastric nitrosation and other intragastric factors including pH, bacterial growth and NO_2 concentration. The analysis of NOC in gastric juice has also presented problems. Walters et al. [25] were the first to develop a method of assaying total NOC as a group in biological materials. By employing this method, several studies confirmed a positive relationship between gastric juice pH, bacterial counts (total and nitrate-reducing), NO_2, and NOC concentrations [26–29]. However, studies employing another method of assaying total NOC [30] gave results with total NOC concentrations approximately 5 times higher than those of Walters et al. [25] and furthermore there appeared to be either no relationship to pH, or even an inverse relationship between intragastric pH and bacterial counts [15, 31–34]. These two methods were critically examined by Pignatelli et al. [35] who showed a lack of reliability of both methods and led to the development by them of a more reliable procedure which did confirm a direct relationship between total gastric juice NOC and pH in one study [35] but an inverse relationship in another [36]. By avoiding the use of nitration stabilizers such as sulphamic acid and assaying fresh gastric juice within a few minutes of its collection, we have been able to increase significantly the sensitivity of the measurements by preventing the decomposition of the unstable NOC especially at high pH, a feature of all the previously published methods [37, 38]. Our studies also showed that significantly higher gastric juice NOC concentrations were found at both a low pH range (1.13–2.99) and high pH range (6.00–8.42) [39]. Thus, in vivo NOC formation can occur through both acid catalysed N-nitrosation (at low pH) as well as in a hypochlorhydric stomach (at a high pH) where it is primarily dependent on the activity of intragastric bacterial flora [24, 40].

Reference has already been made to the inhibiting effect on in vivo N-nitrosation by the concurrent administration of vitamin C, and that many other compounds can also act as catalysts or inhibitors of in vivo and in vitro NOC formation. Sometimes these substances may have both functions depending on the substrate and pH. Vitamins C and E possess antioxidative and other properties that are probably important for cancer prevention. Mirvish et al. [40] first showed that vitamin C could inhibit N-nitrosamine formation at pH 1–4 by competing with secondary amines for nitrosating agents. Mackerness et al. [41] also confirmed that bacterially mediated N-nitrosation is inhibited by vitamin C at high pH. Vitamin E was also found to react with nitrosating agents in both a lipophilic and a hydrophilic medium, but being a more effective inhibitor in the former as it is fat soluble [42, 43]. Ascorbic acid is probably the most effective inhibitor of in vivo N-nitrosation. The underlying mechanism is a reaction between vitamin C and NO_2 to form NO and dehydroascorbic acid and as a consequence vitamin C competes with a nitrosatable amine for the available nitrite [44]. Numerous studies have established vitamin C as a potent inhibitor of specific NOC-induced carcinogenesis in various experimental animals and its inhibitory effect on NOC formation has been reported also in human studies. Animal studies have shown that vitamin C can completely prevent the acute toxic effects attributable to in vivo N-nitrosation in rats and mice gavaged with nitrite and dimethylamine or aminopyrine and tumour induction in rats and mice caused by chronic co-administration of nitrite plus amines or amides can be effectively inhibited by simultaneous administration of vitamin C. Other workers [45] reported that vitamin C inhibited tumour induction in the offspring of pregnant rats gavaged with ethyl urea and nitrite. Ohshima and Bartsch [14] studied the effects of both vitamin C and α-tocopherol (vitamin E) on NPRO formation in rats. NPRO formation inhibition by vitamin C was found to be dose dependent. When 10 times molar excess of ascorbic acid was administered to rats together with precursors, the urinary NPRO excretion in 24 h was suppressed to 2.5% of the control rats.

A similar experiment by the same workers but using vitamin E also resulted in significant though much less marked NPRO reduction. They also investigated the inhibitory effect of these two vitamins on in vivo N-nitrosation in man. Quantitative studies in 1 male volunteer showed that the simultaneous intake of 1 g of vitamin C with the precursors achieved complete inhibition of the N-nitrosation of *L*-proline, the detectable amounts of NPRO being the same as those in controls. On the other hand, the simultaneous intake of 500 mg vitamin E inhibited in vitro N-nitrosation by only about 50% of the effect shown by vitamin C [14]. This difference is predictable as *L*-proline, nitrite and vitamin C are water-soluble compounds compared with vitamin E which is fat soluble. In another study, Wagner et al. [46] also observed only approximately 60% inhibition of endogenous synthesis of NPRO when 400 mg of vitamin E was ingested with precursor. On the other hand, in in vitro studies vitamin E was found to be a very effective inhibitor of N-nitrosopyrrolidine (NPYR) in fried bacon and it also contributed to the reduction of NDMA formation in tobacco smoke [43, 47].

The experimental data on the synergistic influence of ascorbic acid and α-tocopherol both in vitro and in vivo N-nitrosation inhibition are more limited. Merghans et al. [48] in an in vitro study found that a combination of the two vitamins completely blocked the N-nitrosation of aminopyrine at pH 6 and 8. It is of interest that a similar synergistic effect of both these inhibitors was also observed in fried bacon where the inhibition of formed NPYR was greater when the two vitamins were used in combination than when sodium ascorbate was used alone.

An important area of study has been the relevance of gastric factors to gastric carcinogenesis. Studies by Schorah, Sobala and their colleagues [36, 44, 49] in Leeds have shown that ascorbic acid is actively secreted into the gastric lumen and gastric juice concentrations are often much greater than in plasma. However, changes in the gastric mucosa, notably infection with the

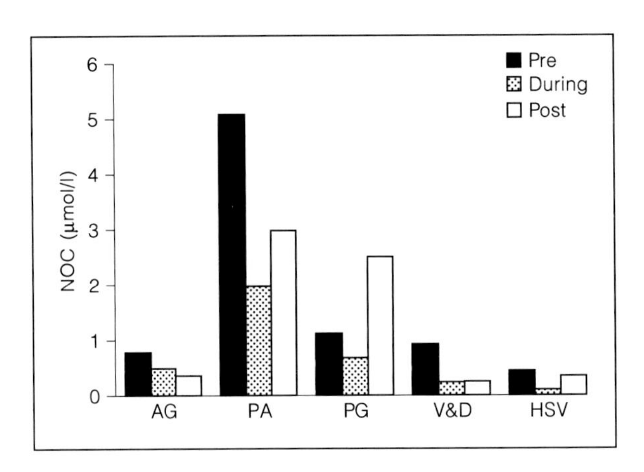

Fig. 1. NOC levels pre-, during and post-vitamin C treatment [53]. AG = Atrophic gastritis; PA = pernicious anaemia; PG = partial gastrectomy; V & D = vagotomy and drainage; HSV = highly selective vagotomy.

organism *Helicobacter pylori,* when associated with chronic gastritis, influence ascorbic acid secretion into the gastric lumen resulting in levels generally lower than those measured concurrently in plasma [49]. Sobala et al. [36] carried out a study in 56 unoperated dyspeptic patients and found no significant differences in plasma vitamin C, gastric NO_2, NO_3 or total NOC concentrations in relation to gastric histology. However, they did find lower ascorbic acid levels in the gastric juice of patients with chronic atrophic gastritis who also had higher pH than normals, though not high enough to allow significant growth of bacteria and thereby elevation in gastric juice NO_2 or NOC levels.

Other have studied specifically the inhibitory effect of vitamin C administration on in vivo N-nitrosation in various patient groups including normal subjects [50], high-risk gastric cancer populations [15, 51–56], high-risk oesophageal cancer populations [34, 8] and high-risk nasopharyngeal carcinoma populations [57]. The vitamin C doses administered in these studies ranged between 75 mg and 4 g daily for periods of 1–28 days.

Reed et al. [53] in England reported the results of administering 4 g vitamin C daily for 4 weeks to 81 subjects at high gastric cancer risk with measurements taken before, at the end of vitamin C treatment and 4 weeks after its discontinuation (fig. 1). Baseline NOC concentrations were elevated in all groups compared with normal controls with a highly significant reduction being observed during treatment ($p<0.001$). Very similar results were seen in NO_2 levels, with significant falls ($p<0.05$) in all high-risk groups except in pernicious anaemia. The results from these studies represent examples of the significant degree to which vitamin C can inhibit endogenous N-nitrosation.

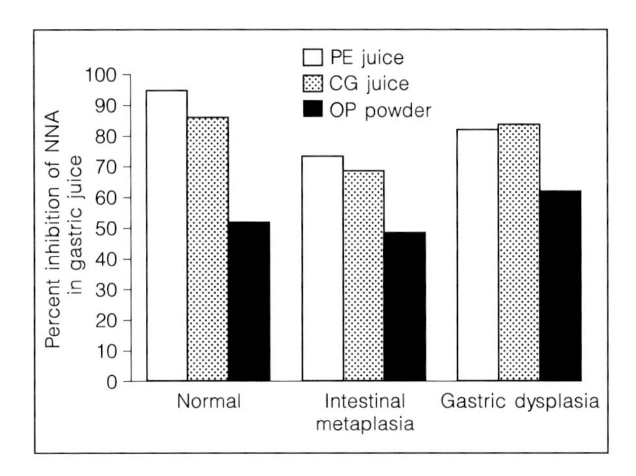

Fig. 2. Effect of *P. emblica* (PE) juice, Chinese gooseberry (CG) juice and orange peel (OP) powder on gastric juice N-nitrosamines (NNA) in man [after 55].

Recently, Xu et al. [55] in China studied the effects of a single-day administration of two high vitamin C content fruits (Chinese gooseberry and *Phyllanthus emblica*), as well as orange peel powder, the dosage of each natural product being adjusted to provide 75 mg vitamin C supplementation in subjects from a high gastric cancer risk area of China (fig. 2). The inhibitory effects of these naturally occurring substances on endogenous gastric juice N-nitrosamine formation are shown in figure 2.

The conclusions to be drawn from all these studies strongly suggest that NOC are important factors in human carcinogenesis. That in vivo intragastric N-nitrosation occurs in man, both acid catalysed at low pH and bacterially catalysed at high gastric pH, has now been conclusively confirmed [56]. However, an indication of the true concentration of NOC formed may not be gained even by employing an improved group-selective total NOC assay procedure. Furthermore, individual characterization of specific NOC, especially those which may be most important in carcinogenesis, has not as yet been achieved. Although it is not known whether the recorded increases in NOC concentrations in the published studies reflect an increase in carcinogenically important NOC, one cannot overlook the fact that the highest gastric juice NOC concentrations were observed in gastric cancer cases in a recently reported study [56]. The development by Pignatelli et al. [58] of an improved method to separate and detect hitherto unknown NOC in biological fluids might make it possible to establish which of these compounds could play an important role in human carcinogenesis. While there are many factors which determine potential carcinogenicity in any individual and the clinical relevance and magnitude of risk of the NOC as potential carcinogens in man needs to be fully

Table 2. Intervention studies – completed and in progress

Authors	Year		Country	Intervention agents	Length of Rx years	Subjects
Completed studies						
Blot et al. [59]	1993	Cardia of stomach	China	Retinol/zinc Riboflavin/niacin Vitamin C/molybdenum β-Carotene/vitamin E/ selenium	5	29,584
Li et al. [60]	1984	Esophageal dysplasia	China	14 vitamins 12 minerals	6	3,318
Current studies						
Correa, pers. commun.	1994	Gastric intestinal metaplasia	Colombia	Partial *H. pylori* eradication Vitamin C and β-carotene	3	700
Muñoz et al. [62]	1993	Precancerous lesions of the stomach	Venezuela	Vitamins C, E and β-carotene	3	3,000
Reed and Johnston [63]	1994	Gastric intestinal metaplasia	Ten European countries	*H. pylori* eradication Vitamin C	3	1,500

determined, the circumstantial evidence is very strong that NOC do play a pivotal role in the aetiology of several human cancers.

It is now well established that certain vitamins do inhibit NOC formation. The wealth of published data from different countries strongly support the importance of vitamins C and E as inhibiting agents of in vivo N-nitrosation and thereby of potential mutagenicity and/or carcinogenicity of many of these NOC.

To confirm the efficacy of these vitamins and other micronutrients as cancer-inhibiting agents, at least of the intestinal type gastric cancer, several formal intervention trials are now in progress in various parts of the world (table 2). It is hoped that they will confirm the protective effects of vitamin supplementation in these specific instances. However, the outcome is unpredictable bearing in mind the results of intervention trials in other areas carried out in China [59, 60] and Finland [61].

Acknowledgements

I wish to thank Belinda Johnston and Jackie Adkins for their assistance in the preparation of the manuscript. Also Dr. Reto Muggli and colleagues at Hoffmann-La Roche, Basel, for supporting several of our vitamin-related studies over a long period of time.

References

1 Bogovski P, Bogovski S: Animal species in which N-nitroso compounds induce cancer. Special report. Int J Cancer 1981;27:471–474.
2 Schmähl D, Scherf HR: Carcinogenic activity of N-nitrosodiethylamine in snakes (*Python reticulatus* Schneider); in O'Neill IK, Von Borstel RG, Miller CT, Long J, Bartsch H (eds): N-Nitroso Compounds: Occurrence, Biological Effects and Relevance to Human Cancer. IARC Sci Publ No 57. Lyon, IARC, 1984, pp 677–682.
3 Preussmann R, Stewart BW: N-nitroso carcinogens; in Searle CE (ed): Chemical Carcinogens. ACS Monogr Ser No 182. Washington, D.C. American Chemical Society, 1984, pp 643–828.
4 Brunnermann KD, Rivenson A, Adams JD, Hecht SS, Hoffmann D: A study of snuff carcinogens; in Bartsch H, O'Neill IK, Schulte-Hermann R (eds): The Relevance of N-Nitroso Compounds to Human Cancer: Exposures and Mechanisms. IARC Sci Publ No 84. Lyon, IARC, 1987, pp 456–459.
5 Hoffmann D, Brunnemann KD, Venitt S: Carcinogenic nitrosamines in oral snuff. Lancet 1988;i: 1232.
6 Anderson RA, Burton HR, Fleming PD, Hamilton-Kemp TR, Gay SL: Effects of air-curing environment on alkaloid-derived nitrosamines in Burley tobacco; in Bartsch H, O'Neill IK, Schulte-Hermann R (eds): The Relevance of N-Nitroso Compounds to Human Cancer: Exposures and Mechanisms. IARC Sci Publ No 84. Lyon, IARC, 1987, pp 451–455.
7 Venitt S: Oral snuff and oral cancer: A new health threat in the UK. Cancer Top 1987;6:75–76.
8 Wu Y, Chen J, Ohshima H, Pignatelli B, Boreham J, Li J, Campbell TC, Peto R, Bartsch H: Geographic association between urinary excretion of N-nitroso compounds and oesophageal cancer mortality in China. Int J Cancer 1993;54:713–719.
9 Hicks RM, Walters CL, Elsebai I, El Asser A-B, El Merzatani M, Gough TA: Demonstration of nitrosamines in human urine. Preliminary observations on a possible aetiology for bladder cancer in association with chronic urinary tract infections. Proc R Soc Med 1977;70:413.
10 Shephard SE, Meier I, Lutz WK: Alkylating potency of nitrosated amino acids and peptides; in O'Neill IK, Chen J, Bartsch H (eds): Relevance to Human Cancer of N-Nitroso Compounds, Tobacco Smoke and Mycotoxins. IARC Sci Publ No 105. Lyon, IARC, 1991, pp 383–387.
11 Hecht SS, Hoffmann D: N-nitroso compounds and tobacco-induced cancers in man; in O'Neill IK, Chen J, Bartsch H (eds): Relevance to Human Cancer of N-Nitroso Compounds, Tobacco Smoke and Mycotoxins. IARC Sci Publ No 105. Lyon, IARC, 1991, pp 54–61.
12 Druckrey H, Preussmann R, Ivankovic S, Schmähl D: Organotropic carcinogenicity of 65 different N-nitroso compounds in BD rats. Z Krebsforsch 1967;69:103–210.
13 Sander J, Bürkle G: Induction of malignant tumours in rats by simultaneous feeding of nitrite and secondary amines. Z Krebsforsch 1969;76:93–96.
14 Ohshima H, Bartsch H: Quantitative estimation of endogenous nitrosation in humans by monitoring N-nitrosoproline excreted in the urine. Cancer Res 1981;41:3658–3662.
15 Hall CN, Kirkham JS, Northfield TC: Urinary N-nitrosoproline excretion: A further evaluation of the nitrosamine hypothesis of gastric carcinogenesis in precancerous conditions. Gut 1987;28:218.
16 Adam B, Schlag P, Friede P, Preussmann R, Eisenbrand G: Proline is not useful as a chemical probe to measure nitrosation in the gastrointestinal tract of patients with gastric disorders characterised by anacidic conditions. Gut 1989;30:1068–1075.
17 Leach SA, Challis B, Cook AR, Hill MJ, Thompson MH: Bacterial catalysis of the N-nitrosation of secondary amines. Biochem Soc Trans 1985;13:380–381.

18 Tannenbaum SR: Endogenous formation of N-nitroso compounds: A current perspective; in Bartsch H, O'Neill IK, Schulte-Hermann R (eds): The Relevance of N-Nitroso Compounds to Human Cancer: Exposures and Mechanisms. IARC Sci Publ No 84. Lyon, IARC, 1987, pp 292–296.
19 Lu SH, Yang WX, Guo LP, Li FM, Wang GJ, Zhang JS, Li PZ: Determination of N-nitrosamines in gastric juice and urine and a comparison of endogenous formation of N-nitrosoproline and its inhibition in subjects from high-risk and low-risk areas for oesophageal cancer; in Bartsch H, O'Neill IK, Schulte-Hermann R (eds): The Relevance of N-Nitroso Compounds to Human Cancer: Exposures and Mechanisms. IARC Sci Publ No 84. Lyon, IARC, 1987, pp 537–543.
20 Mirvish SS, Issenberg P, Sams JP: A study of N-nitrosomorpholine synthesis in rodents exposed to nitrogen dioxide and morpholine; in Scanlan RA, Tannenbaum SR (eds): ACS Monogr Ser No 174. Washington, 7 D.C. American Chemical Society, 1981, pp 181–191.
21 Garland WA, Kuenzig W, Rubio F, Kornchuk H, Norkus EP, Conney AH: Urinary excretion of nitrosodimethylamine and nitrosoproline in humans: Interindividual and intraindividual differences and the effect of administered ascorbic acid. Cancer Res 1986;46:5392–5400.
22 Correa P, Haenszel W, Cuello C, Tannenbaum S, Archer M: A model for gastric cancer epidemiology. Lancet 1975;ii:58–60.
23 Leach SA, Cook AR, Challis BC, Hill MJ, Thompson MH: Bacterially mediated N-nitrosation reactions and endogenous formation of N-nitroso compounds; in Bartsch H, O'Neill IK, Schulte-Hermann R (eds): The Relevance of N-Nitroso Compounds to Human Cancer: Exposures and Mechanisms. IARC Sci Publ No 84. Lyon, IARC, 1987, pp 396–399.
24 Leach SA: Mechanisms of endogenous N-Nitrosation; in Hill MJ (ed): Nitrosamines: Toxicology and Microbiology. Chichester, Ellis Horwood, 1988, pp 69–87.
25 Walters CL, Downes MJ, Edwards MW, Smith PLR: Determination of a non-volatile N-nitrosamine on a food matrix. Analyst 1978;103:1127–1133.
26 Reed PI, Haines K, Smith PLR, Walters CL, House FR: The gastric juice N-nitrosamines in health and gastro-duodenal disease. Lancet 1981;ii:550–552.
27 Walters CL, Smith PLR, Reed PI, Haines K, House FR: N-nitroso compounds in gastric juice and their relationship to gastroduodenal disease. IARC Sci Publ No 41. Lyon, IARC, 1982, pp 345–352.
28 Stockbrugger RW, Cotton PB, Eugenides N, Bartholomew BA, Hill MJ, Walters CL: Intragastric nitrite, nitrosamines and bacterial overgrowth during cimetidine treatment. Gut 1982;23:1048–1052.
29 Sharma BK, Santana IA, Wood EC, Walt RP, Peraira M, Noone P, Smith PLR, Walters CL, Pounder RE: Intragastric bacterial activity and nitrosation before, during, and after treatment with omeprazole. Br Med J 1984;289:717–719.
30 Bavin PMG, Darkin DW, Viney NJ: Total nitroso compounds in gastric juice; in Bartsch H, O'Neill IK, Castegnaro M, Okada M (eds): N-Nitroso Compounds: Occurrence and Biological Effects. IARC Sci Publ No 41. Lyon, IARC, 1981, pp 337–343.
31 Milton-Thompson GJ, Lightfoot NF, Ahmet Z, Hunt RH, Barnard J, Bavin PMG, Darkin DW, Vivey N: Intragastric acidity, bacteria, nitrite, and N-nitroso compounds before, during and after cimetidine treatment. Lancet 1982;i:1091–1095.
32 Keighley M, Youngs D, Poxon V, Morris D, Muscroft TJ, Burdon DW, Barnard J, Bavin PMG, Brimblecombe RW, Darkin DW, Moore PJ, Viney N: Intragastric nitrosation is unlikely to be responsible for gastric carcinoma developing after operation for duodenal ulcer. Gut 1984;25: 238–245.
33 Hall CN, Darkin D, Brimblecombe R, Cook AJ, Kirkham JS, Northfield TC: Evaluation of the nitrosamine hypothesis of gastric carcinogenesis in precancerous conditions. Gut 1986;27:491–498.
34 Crespi M, Ohshima H, Ramazzotti V, Muñoz N, Grassi A, Casala V, Lechera H, Calmels S, Cattoen C, Kaldor J, Bartsch H: Intragastric nitrosation and precancerous lesions of the gastrointestinal tract: Testing of an aetiological hypothesis; in Bartsch H, O'Neill IK, Schulte-Hermann R (eds): The Relevance of N-Nitroso Compounds to Human Cancer: Exposures and Mechanisms. IARC Sci Publ No 84. Lyon, IARC, 1987, pp 511–517.
35 Pignatelli B, Richard I, Bourgade MC, Bartsch H: Improved group determination of total N-nitroso compounds in human gastric juice by chemical denitrosation and thermal energy analysis. Analyst 1987;112:945–949.

36 Sobala GM, Pignatelli B, Shorah CJ, Bartsch H, Sanderson M, Dixon MF, Shires S, King RFG, Axon ATR: N-nitroso compounds, ascorbic acid and total bile acids in gastric juice of patients with and without precancerous conditions of the stomach. Carcinogenesis 1991;12:193–198.
37 Xu GP, Reed PI: A method for group determination of total N-nitroso compounds and nitrite in fresh gastric juice by chemical denitrosation and thermal energy analysis. Analyst 1993;118:877–883.
38 Xu GP, Reed PI: Instability of N-nitroso compounds in gastric juice and preliminary results from analyses of fresh samples by using an improved analytical method. Eur J Cancer Prev 1993;2: 381–386.
39 Xu GP, Reed PI: N-nitroso compounds in fresh gastric juice and their relation to intragastric pH and nitrite employing an improved analytical method. Carcinogenesis 1993;14:2547–2551.
40 Mirvish SS, Wallcave L, Eager M, Shubik P: Ascorbate-nitrite reaction: Possible means of blocking the formation of carcinogenic N-nitroso compounds. Science 1972;177:65–68.
41 Mackerness CW, Leach SA, Thompson MH, Hill MJ: Inhibition of bacterially mediated N-nitrosation by vitamin C: Relevance to the inhibition of endogenous N-nitrosation in the achlorhydric stomach. Carcinogenesis 1989;10:397–399.
42 Lathia D, Braasch A, Theissen I: Inhibitory effects of vitamin C and E on in-vitro formation of N-nitrosamine under physiological conditions. Gastrointest Res 1988;14:151–156.
43 Mirvish SS: Effect of vitamins C and E on N-nitroso formation, carcinogenesis and cancer. Cancer 1986;58:1842–1850.
44 Sobala GM, Schorah CJ, Sanderson M, Dixon MF, Tompkins DS, Godwin P, Axon ATR: Ascorbic acid in the human stomach. Gastroenterology 1989;97:357–363.
45 Ivankovic S, Preussmann R, Schmähl D, Zeller WJ: Prevention by ascorbic acid of in vivo formation of N-nitroso compounds; in Bogovski P, Walker EA (eds): N-Nitroso Compounds in the Environment. IARC Sci Publ No 9. Lyon, IARC, 1974, pp 101–102.
46 Wagner DA, Shuker DEG, Bilmazes C, Obdiedzinski M, Baker I, Young RV, Tannenbaum SR: Effect of vitamins C and E on endogenous synthesis of N-nitrosamino acids in humans: Precursor-product studies with (^{15}N)nitrate. Cancer Res 1985;45:6519–6522.
47 Mergans WJ, Kamm JJ, Newmark HL: Alpha-tocopherol. Uses in preventing nitrosamine formation. IARC Sci Publ No 19. Lyon, IRAC, 1978, pp. 199–221.
48 Merghans WJ, Chau J, Newmark HL: The influence of ascorbic acid and *D*-alpha-tocopherol on the formation of nitrosamines in an in vitro gastrointestinal model system. IARC Sci Publ No 31. Lyon, IARC, 1980, pp 259–266.
49 Schorah CJ, Sobala GM, Sanderson M, Collis N, Primrose JN: Gastric juice ascorbic acid: Effects of disease and implications for gastric carcinogenesis. Am J Clin Nutr 1991;53:287S–293S.
50 Bartsch H, Ohshima H, Pignatelli B: Inhibition of endogenous nitrosation: Mechanisms and implications in human cancer prevention. Mutat Res 1988;202:307–324.
51 Reed PI, Summers K, Smith PLR, Walters CL, Bartholomew B, Hill MJ, Venitt S, Hornig D, Bonjour J-P: Effect of ascorbic acid treatment on gastric juice nitrite and N-nitroso compound concentration in achlorhydric subjects. Gut 1983;24:A492–A493.
52 Kamiyama S, Ohshima H, Shimada A, Siato N, Bourgade M-C, Ziegler P, Bartsch H: Urinary excretion of N-nitrosamino acids and nitrate by inhabitants in high- and low-risk areas for stomach cancer in northern Japan; in Bartsch H, O'Neill IK, Schulte-Hermann R (eds): The Relevance of N-Nitroso Compounds to Human Cancer: Exposures and Mechanisms. IARC Sci Publ No 84. Lyon, IARC, 1987, pp 497–502.
53 Reed PI, Johnston BJ, Walters CL, Hill MJ: Effect of ascorbic acid on the intragastric environment in patients at increased risk of developing gastric cancer; in O'Neill IK, Chen J, Bartsch H (eds): Relevance to Human Cancer of N-Nitroso Compounds, Tobacco Smoke and Mycotoxins. IARC Sci Publ No 105. Lyon, IARC, 1991, pp 139–142.
54 Sierra R, Ohshima H, Muñoz H, Teuchmann S, Rana AS, Laveille C, Pignatelli B, Chinnock A, Ghissazzi F El, Chan C, Hautefeuille A, Gamboa C, Bartsch H: Exposure to N-nitrosamines and other risk factors for gastric cancer in Costa Rican children; in O'Neill IK, Chen J, Bartsch H (eds): Relevance to Human Cancer of N-Nitroso Compounds, Tobacco Smoke and Mycotoxins. IARC Sci Publ No 105. Lyon, IARC, 1991, pp 162–167.

55 Xu GP, Song PJ, Reed PI: Hypothesis on the relationship between gastric cancer and intragastric nitrosation: N-nitrosamines in gastric juice of subjects from a high-risk area for gastric cancer and the inhibition of N-nitrosamine formation by fruit juices. Eur J Cancer Prev 1993;2:25–36.
56 Reed PI, Xu GP, Li DH, Hill MJ, Johnston BJ: N-nitroso compound levels in fresh gastric juice in relation to clinical diagnosis. Gut 1993;34(suppl 4):530.
57 Zeung Y, Ohshima H, Bouvier G, Roy P, Jianming Z, Li B, Bionet I, de Thé G, Bartsch H: Urinary excretion of nitrosamino acids and nitrate by inhabitants of high-risk and low-risk areas for nasopharyngeal carcinoma in Southern China. Cancer Epidemiol Biomarkers Prev 1993;2: 195–200.
58 Pignatelli B, Malaveille C, Therillier P, Hautefeuille A, Bartsch H: Improved methods for analysis of N-nitroso compounds and applications in human biomonitoring; in Loeppky RN, Michejda CJ (eds): Nitrosamines and Related N-Nitroso Compounds. Chemistry and Biochemistry. ACS Symp Ser No 553. Washington, 7 D.C. American Chemical Society, 1992, pp 102–118.
59 Blot WJ, Li JY, Taylor PR, Guo W, Dawsey S, Wang GQ, Yang CQ, Chung S, Zheng SF, Gail M, Li GY, Yu Y, Liu BQ, Tangrea J, Sun YH, Liu F, Fraumeni JF Jr, Zhang YH, Li B: Nutrition intervention trials in Linxian, China: Supplementation with specific vitamin/mineral combinations, cancer incidence, and disease-specific mortality in the general population. J Natl Cancer Inst 1993; 85:1483–1491.
60 Li JY, Taylor PR, Li B, Dawsey S, Wang GQ, Ershow AG, Guo W, Liu SF, Yang CS, Shen Q, Wang W, Mark SD, Zou XN, Greenwald P, Wu YP, Blot WJ: Nutrition intervention trials in Linxian, China: Multiple vitamin/mineral supplementation, cancer incidence and disease-specific mortality among adults with esophageal dysplasia. J Natl Cancer Inst 1993;85:1492–1498.
61 The Alpha-Tocopherol Beta-Carotene Cancer Prevention Study Group: The effect of vitamin E and beta-carotene on the incidence of lung cancer and other cancers in male smokers. N Engl Med J 1994;330:1029–1035.
62 Muñoz N, Vivas J, Buiatti E, Peraza S, de Sanjosé S, Carno E, Castro D, Sanchez V, Andrade O, Benz M, Oliver W: Chemoprevention trial of precancerous lesions of the stomach in Venezuela. Eur J Cancer Prev 1993;2(suppl 1):5.
63 Reed PI, Johnston BJ: Primary prevention of gastric cancer – The ECP-IM Intervention Trial. Acta Endoscop 1995;25:45–54.

Dr. Peter I. Reed, Lady Sobell Gastrointestinal Unit, Wexham Park Hospital,
Slough, Berks SL2 4HL (UK)

Walter P (ed): The Scientific Basis for Vitamin Intake in Human Nutrition.
Bibl Nutr Dieta. Basel, Karger, 1995, No 52, pp 43–55

Interaction of Vitamins with Mental Performance[1]

Helmut Heseker[a], *Werner Kübler*[b], *Volker Pudel*[c], *Joachim Westenhöfer*[c]

[a] Nutrition Department, University of Paderborn;
[b] Institute of Nutritional Sciences, University of Giessen, and
[c] Research Group for Nutritional Psychology, University of Göttingen, Germany

For several vitamins, neurological and/or behavioral impairment is associated with overt deficiency. In severe stages of beriberi or pellagra, typical psychosyndromes can be diagnosed. Fortunately these stages of vitamin deficiencies can rarely be found in industrialized countries but are not uncommon in developing countries [1]. It is well documented from vitamin deprivation studies, that alterations in behavior and mental performance arise already in early stages of vitamin deficiencies, which might be clinically relevant [2–6]. Mild to moderate vitamin deficiencies are possibly much more common in our populations [7]. In correlational studies, evaluating the association between vitamin status and neurocognitive function in healthy elderly people, subjects with low vitamin C, B_{12}, folate, thiamin or riboflavin status scored poorly on tests of memory, nonverbal thinking or behavior [8–10]. Despite the limitations of such correlational data, the implication that poor vitamin status can contribute to a decline of neurocognitive function in otherwise healthy persons is challenging.

The objective of our study was to investigate the functional significance of mild to moderate vitamin deficiencies in a large group of free-living volunteers. The association of vitamin status and psychological function was examined in a random, double-blind trial with an 8-week supplementation period in healthy young men.

[1] This study was supported by the German Ministry for Research and Technology, grant numbers 0 704 722 3 and 0 704 724 5. The responsibility for the contents of this publication lies with the authors.

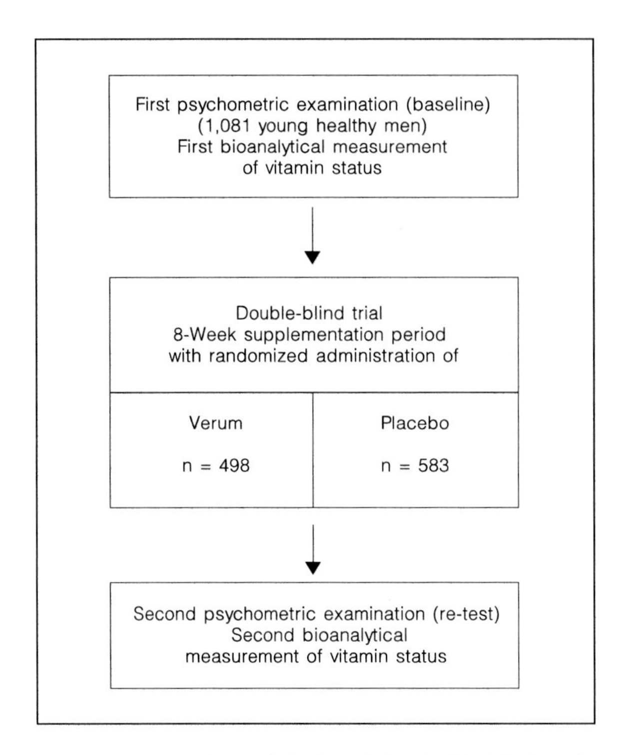

Fig. 1. Experimental design of the double-blind study.

Materials and Methods

In a placebo-controlled intervention study (1,081 healthy young men, aged 17–29 years) the relation between the vitamin status and behavior and mental performance – determined by psychometric measurements – was studied (fig. 1). The vitamin status was determined by bioanalytical measurements, regarding the vitamins A, E, C, thiamin, riboflavin, B_6, folate and cobalamin (table 1). To record the psychometric variables, a computerized test battery was set up, which included tests in different diagnostic areas such as mental performance capacity, enduring personality traits as well as the current mental status (table 2). To minimize external influences, the test situation was standardized as much as possible. A more detailed description of the used methods can be found somewhere else [11, 12]. The baseline examination was followed by an 8-week supplementation period (verum/placebo), before the measurements were repeated. The supplement contained vitamins in a physiological range (table 3). Verum and placebo groups did not differ in their demographic, biochemical and psychometric variables at the beginning of the study.

Table 1. Vitamin parameters and analytical methods [12]

Vitamin	Parameter	Method
Vitamin C	plasma-ascorbic acid	DNPH; photometrical
Folate	plasma-folate	radioassay[1]
Vitamin B_{12}	plasma-cobalamin	radioassay[1]
Vitamin A	plasma-retinol	HPLC
Vitamin E	plasma-tocopherol	HPLC
Thiamin	α-ETK	enzyme assay
Riboflavin	α-EGR	enzyme assay
Vitamin B_6	α-EAST	enzyme assay

[1] Radioassay kit from Clinical Assays.

Results and Discussion

The results of this clinical trial allow a statistical evaluation of different questions:

1. Do vitamin-deficient and control groups – classified by the first bioanalytical measurement of the vitamin supply – differ in their psychometric values at the beginning of the study?

To answer this question, low-vitamin-status and control groups were created for each tested vitamin by evaluating the first bioanalytical measurement. Persons with an observed vitamin value below the 5th percentile (above the 95th percentile for activity coefficients) were defined as 'deficient'. In the control group all vitamin supply values were above the median (below the median for activity coefficients). After this procedure the control and 'deficient' groups consisted of roughly 50 persons.

The following statistical analysis of the psychometric measurements revealed significant detrimental results only in a small number of the psychometric scales in the low-vitamin-status groups. The number of plausible and significant differences did not exceed the statistically expected 5% significant results. A relationship between vitamin status and the tested psychometric scales could not be established in this evaluation step. A single determination of the vitamin supply is possibly not sufficient to describe reliably the real, long-term vitamin status of an individual. However, when no long-term parameters for evaluating vitamin supply are available, only repeated bioanalytical monitoring of vitamin supply can adequately confirm a chronic vitamin deficiency.

Table 2. Psychometric tests [12]

Test procedure	Diagnostic area	Parameter
Vienna test system (VTS)	psychological performance	reaction time long-term attentiveness concentration vigilance susceptibility to stress visual-motor coordination
Freiburg personality inventory (FPI)	personality dimensions	nervousness aggressiveness depressiveness irritability sociability lassitude domineering inhibition frankness extroversion emotional lability masculinity
Adjective check-list (ACL)	current mental state	activity ability to concentrate inactivity fatigue drowsiness extroversion introversion self-confidence heightened mood excitation sensitivity anger anxiety depression dreaminess

Table 3. Mean vitamin intake of the volunteers before the 8-week supplementation period and from the supplement

Vitamin	Mean daily vitamin intake from food	Supplement
Thiamin, mg	1.4	3.0
Riboflavin, mg	1.7	3.5
Vitamin B_6, mg	1.8	4.5
Niacin[1], mg	14	20
Folate, μg	214	800
Vitamin B_{12}, μg	6.4	6.0
Vitamin C, mg	87	150
Vitamin A[2], mg	1.0	1.5
Vitamin E, mg	15.4	20.0

[1] Niacin equivalents.
[2] Retinol equivalents.

2. Do vitamin-deficient and control groups – as classified by the repeated bioanalytical measurement of the vitamin status – differ in their psychometric values at the beginning of the study?

Effects of a deficient vitamin supply on physiological and psychological functions are expected mainly in cases where there is a long-term inadequate vitamin supply and less so in cases where there is only a short-term and temporary deficit. Assessing vitamin status solely on the basis of the data obtained at the beginning of the experiment may lead to misclassifications, because the individual vitamin supply can be subject to considerable fluctuations. The consequence is a poor discrimination between the postulated deficient and control groups as shown in figures 2a,b and 3a,b.

In contrast to the vitamin-treated group, the intake of vitamins in the placebo group was not changed by experimental conditions. By comparing the vitamin supply parameters of this group at the beginning of the study with the repeated measurement it was possible to differentiate subgroups with a stable, poor supply of certain vitamins. Classification into control and deficient group again was done for each vitamin separately. Through this strict selection procedure, subpopulations with chronic poor vitamin supply were compared with subpopulations with a stable sufficient vitamin supply. Bioanalytical vitamin supply values, which are associated with an increasing number of unfavorable psychometric findings, are shown in table 4.

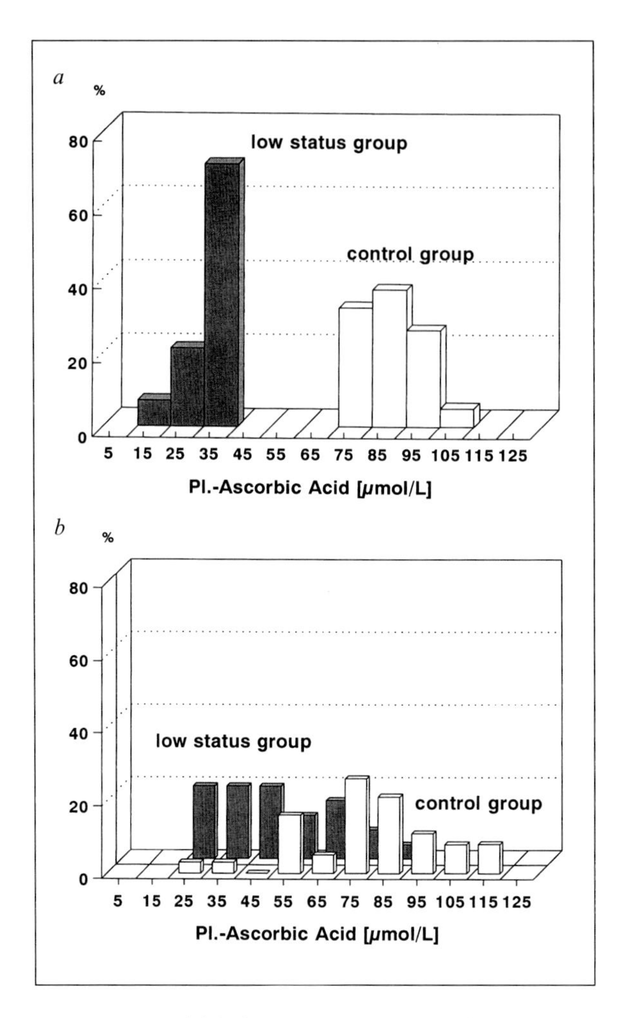

Fig. 2. a Initial plasma concentrations of ascorbic acid in the low-vitamin-status and control group (placebo group, n = 50, first measurement). *b* Plasma concentrations of ascorbic acid in the low-vitamin-status and control group (same persons as in figure 2a, second measurement).

A general review of the results show that current mental status (ACL scales) is often more influenced than long-term components of behavior (FPI scales) (fig. 4). The ACL scales show that poor vitamin supplies are accompanied by a decreased sense of well-being, a heightened emotional irritability and an increased feeling of fear. The FPI scales reveal that a vitamin deficit is accompanied principally by heightened nervousness, depression and more extraversion. The different subtests of the Vienna test system revealed signific-

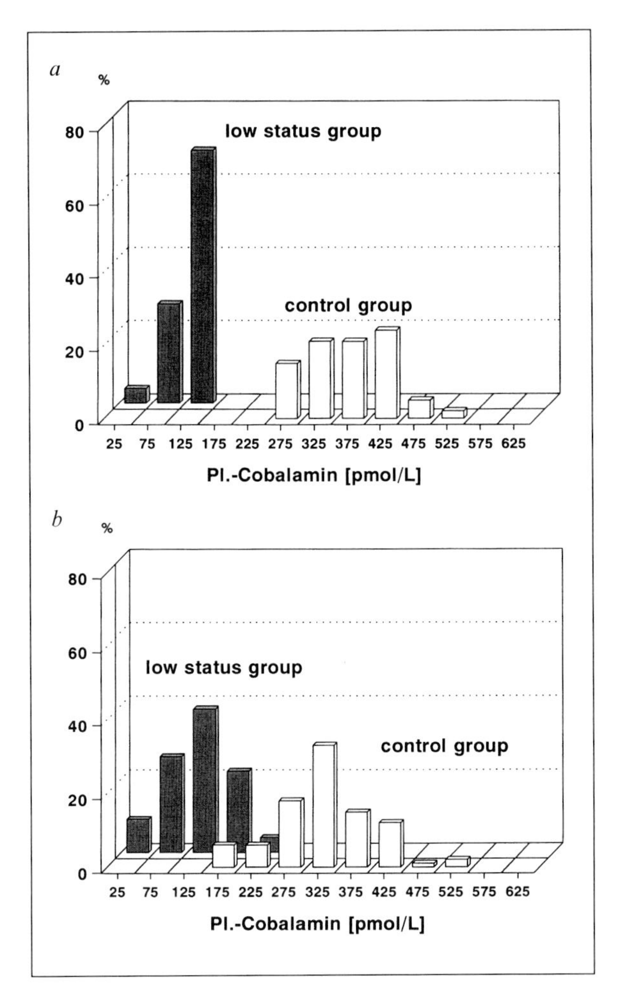

Fig. 3. a Initial plasma concentrations of vitamin B_{12} in the low-vitamin-status and control group (placebo group, n = 50, first measurement). *b* Plasma concentrations of vitamin B_{12} in the low-vitamin-status and control group (same persons as in figure 3a, second measurement).

ant poor memory and reaction performance in thiamin-deficient groups. For vitamin B_6 and A overall only few unfavorable results were found. The large number of significant differences which greatly exceed the expected statistical value of 5% and also the plausibility of the results suggest that there is an interaction between vitamin status and mental performance in healthy people. This analysis indicates that functional disorders can occur even with a slightly unfavorable vitamin status if the deficiency has existed for a long period of

Table 4. Bioanalytical vitamin supply values associated with an increasing number of unfavorable psychometric findings

Parameter	Cut-off value
Plasma-ascorbic acid	≤ 50 μmol/l
Plasma-folate	≤ 10 nmol/l
Plasma-cobalamin	≤ 170 pmol/l
Plasma-tocopherol	≤ 20 μmol/l
α-ETK	≥ 1.16
α-EGR	≥ 1.50
α-EAST	≥ 1.80

time. This seems plausible since it cannot be assumed that a short-term deficit in a vitamin supply leads to deficiency symptoms. The results from this study are consistent with the results from a former study on free-living, almost healthy elderly [8]. Nevertheless, correlational data have some limitations and sometimes several explanations for the observed relationships are possible: neurological and psychological disorders may lead to poor dietary practices and consequently vitamin deficiencies [13]. A causal relationship can only be established by improving the vitamin supply. Therefore a placebo-controlled intervention study followed this baseline examination.

3. Does vitamin supplementation improve psychometric values per se as compared with a placebo treatment, without taking the initial vitamin status into account?

The vitamin supplementation of the verum group results in an increase of the vitamin concentrations in plasma (exception: retinol) and a decrease of the activity coefficients (table 5). The observed different effects on the vitamin status cannot be explained only by different supplementation levels, but rate of absorption, kinetic and elimination properties also determine the quantity of the increase of the steady-state concentrations in plasma and tissues [14].

Possible advantages of the 8-week multivitamin supplementation versus administration of placebo, initially with no regard to vitamin status at the outset of the trial, were statistically tested (two-way analysis of variance). No significant differences were found in any of the psychometric parameters. This supports the important conclusion that vitamin supplementation in physiological dosages above normal levels usually obtained from food sources, does not lead to any measurable improvement in the psychometric parameters studied

Table 5. Effect of vitamin supplementation on the vitamin status (multivitamin group, n = 498)

Vitamin parameter	Before supplementation			After supplementation			p[1]
	$\bar{x}$	SEM	median	$\bar{x}$	SEM	median	
Plasma vitamin concentrations							
Ascorbic acid, μmol/l	70.6	0.83	71.0	83.1	0.75	82.9	<0.001
Folate, nmol/l	15.8	0.24	14.7	34.0	0.62	32.0	<0.001
Cobalamin, pmol/l	293	5.5	275	345	5.9	321	<0.001
Retinol, μmol/l	1.93	0.02	1.92	1.89	0.02	1.85	n.s.[2]
α-Tocopherol, μmol/l	26.0	0.25	25.3	29.0	0.32	27.8	<0.001
Enzyme assays[3]							
α-ETK	1.11	0.002	1.11	1.07	0.002	1.07	<0.001
α-EGR	1.35	0.006	1.34	1.18	0.003	1.17	<0.001
α-EAST	1.58	0.006	1.57	1.50	0.006	1.49	<0.001

[1] t test, significantly different if $p \leq 0.05$.
[2] Not significant.
[3] Enzyme activities related to pure erythrocytes (hematocrit = 1.0).

here, given that no previous vitamin deficiency existed. The vitamin concentration in the brain is controlled by an effective regulatory system at the brain-blood barrier [15]. Therefore, moderately increased vitamin levels in blood will possibly not result in elevated vitamin concentration in the cerebrospinal fluid or the brain.

4. Does vitamin supplementation improve psychometric values as compared with a placebo treatment, if a mild to moderate vitamin deficiency exists at the beginning of the study?

To determine whether vitamin supplementation in volunteers with an insufficient supply of vitamins results in a significant improvement, the psychometric results between the active and placebo groups were compared before and after supplementation. Deficient placebo and deficient verum groups were formed using the 5th percentile (95th percentile for the activity coefficients) as the cut-off value. Effects of the vitamin supplementation were observed for folic acid, vitamin C and, to a lesser extent, thiamin. Administration of the vitamins, as compared to placebos, in men with an initial suboptimal *folate* status led to a decreased emotional lability, increased activity and concentration, higher extraversion and lower introversion, greater self-confidence and a markedly improved mood. In volunteers with an initial mild to moderate

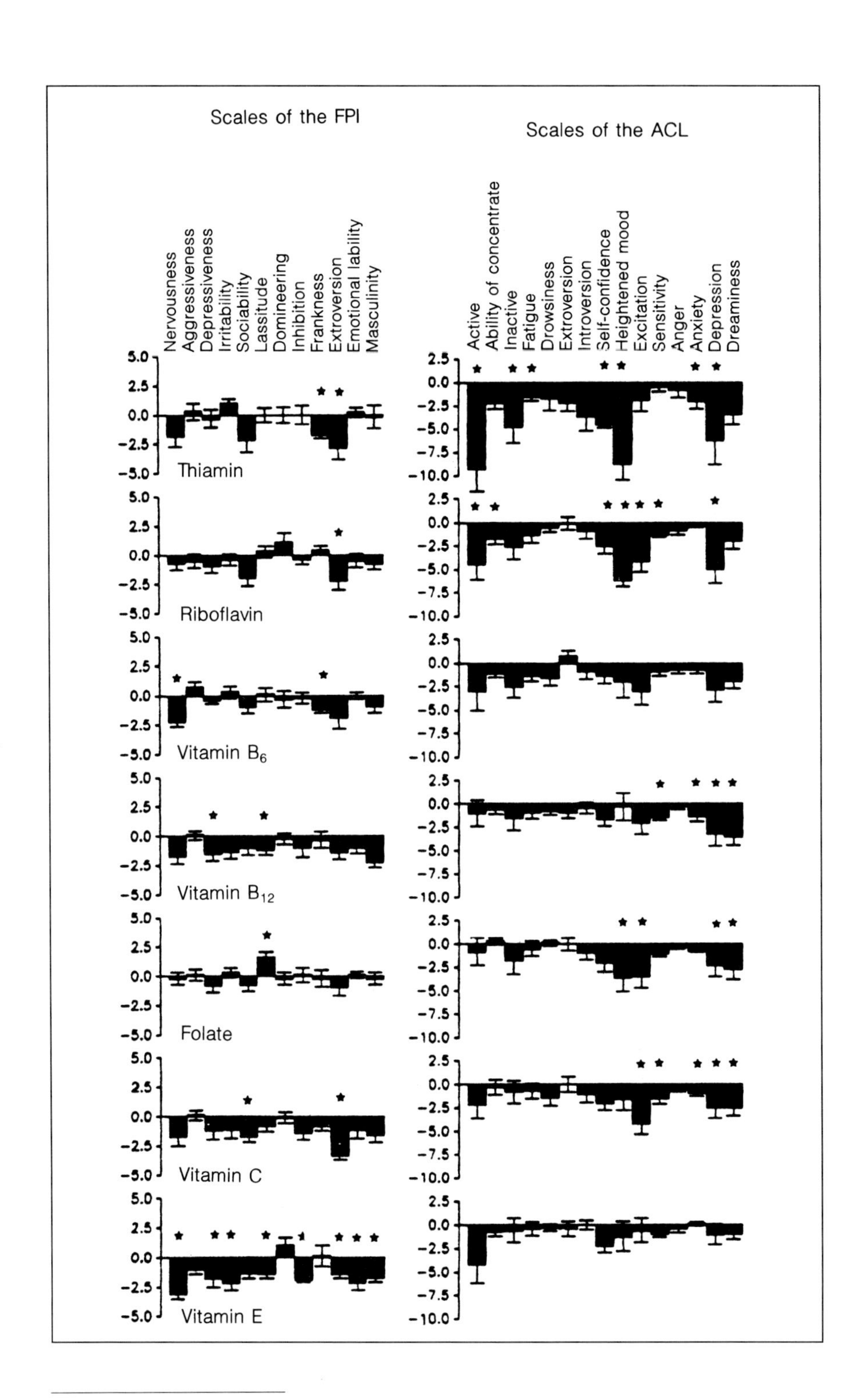
Scales of the FPI
Scales of the ACL
Nervousness
Aggressiveness
Depressiveness
Irritability
Sociability
Lassitude
Domineering
Inhibition
Frankness
Extroversion
Emotional lability
Masculinity
Active
Ability of concentrate
Inactive
Fatigue
Drowsiness
Extroversion
Introversion
Self-confidence
Heightened mood
Excitation
Sensitivity
Anger
Anxiety
Depression
Dreaminess
Thiamin
Riboflavin
Vitamin B6
Vitamin B12
Folate
Vitamin C
Vitamin E

vitamin C deficiency, supplementation led to decreased nervousness, less depression and decreased emotional lability. In men with a low *thiamin* status a statistically positive influence of thiamin on sociability and sensitivity was confirmed. For the other vitamins, the proposed null hypothesis (h_0) – that vitamin supplements had no effect on individual psychometric test scales – could not be disproven.

The simple assessment of vitamin supply included not only volunteers with a chronic deficiency but also those with temporary suboptimal vitamin intake. Probably changes in the vitamin-treated groups are not as pronounced as they would have been if the volunteers had a chronically insufficient vitamin intake. These findings indicate that certain psychometric and bioanalytical vitamin supply parameters not only correlate, but that apparently a cause-effect relationship between the two factors exists. This relationship is possibly much stronger if it would be possible to measure the vitamin status directly in specific organ systems, including the nervous system instead of determining blood concentrations. Concentrations of vitamins or their metabolites in blood may be less predictive to psychometric parameters.

High Dosages of Vitamins

The number of studies testing the effect of physiological vitamin dosages is limited opposed to megavitamin studies. High dosages of water-soluble vitamins have been used to treat various disorders of the central nervous system including schizophrenia.

A review of controlled trials in patients with psychiatric diseases indicates that there is no adequate support for beneficial effects of high dosages of niacin or vitamin B_6, if no vitamin deficiency is manifest [16]. Nevertheless, examples of the benefit of megavitamin therapy in vitamin-dependent inborn errors of metabolism are well known [17].

Oral application of elevated dosages of vitamin B_1 (90 mg/day), B_6 (60 mg/day) and B_{12} (120 μg/day) for example has been found to improve target shooting in marksmen in two placebo-controlled double-blind studies. It could be demonstrated that the improved firing accuracy was closely associated with a significant reduced physiological tremor [18].

Fig. 4. Psychometric data in chronic vitamin deficiency. Shown are the absolute differences compared with an adequately supplied control group. Unfavorable results as compared with control are printed as negative values ($\bar{x} \pm$ SEM). Significant differences (t test; $p \leq 0.05$) are indicated with an asterisk.

Conclusion

In the presence of a chronically insufficient vitamin supply which was verified by repeated measurements of the vitamin parameters, many unfavorable psychometric findings in the corresponding deficiency groups are observed for the vitamins C, thiamin, riboflavin, cobalamin and folate, depending on the degree of the insufficient vitamin supply. Vitamin supplementation in cases of initially insufficient vitamin supply indicate some effects in the sense of an improvement of behavior and cognitive functions. Supplemental vitamin intake in physiological dosages in addition to a vitamin-sufficient diet did not lead to an improvement of behavior and mental performance.

References

1 Djoenaidi W, Notermans SLH, Dunda G: Beriberi cardiomyopathy. Eur J Clin Nutr 1992;46: 227–234.
2 Brozek J: Psychological effects of thiamine restriction and deprivation in normal young men. Am J Clin Nutr 1957;5:109–118.
3 Burvill PW, Jackson JM, Smith WG: Psychiatric symptoms due to vitamin B_{12} deficiency without anemia. Med Aust 1969;2:388–390.
4 Kinsman RH, Hood J: Some behavioral effects of ascorbic acid deficiency. Am J Clin Nutr 1971; 24:455–464.
5 Milner G: Ascorbic acid deficiency in chronic psychiatric patients – A controlled trial. Br J Psychiatry 1963;109:294–299.
6 Sterner RT, Price RW: Restricted riboflavin: Within subject behavioral effects in humans. Am J Clin Nutr 1973;26:150–160.
7 Buzina R, Bates CJ, van der Beek E, Brubacher G, Chandra RK, Hallberg L, Heseker H, Mertz W, Pietrzik K, Pollitt E, Pradilla A, Suboticanec A, Sandstead HH, Schalch W, Spurr GB, Westenhöfer W: Workshop on functional significance of mild-to-moderate malnutrition. Am J Clin Nutr 1989;50:172–176.
8 Chomé J, Paul T, Pudel V, Bleyl H, Heseker H, Hüppe R, Kübler W: Effects of suboptimal vitamin status on behavior. Bibl Nutr Dieta. Basel, Karger, 1986, No 38, pp 94–103.
9 Goodwin JS, Goodwin JM, Garry PJ: Association between nutritional status and cognitive functioning in a healthy elderly population. JAMA 1983;249:2917–2921.
10 Tucker DM, Penland JO, Sandstead HH: Nutrition status and brain function in aging. Am J Clin Nutr 1990;52:93–102.
11 Heseker H, Kübler W, Westenhöfer J, Pudel V: Psychische Veränderungen als Frühzeichen einer suboptimalen Vitaminversorgung. Ernähr Umsch 1990;37:87–94.
12 Heseker H: Zur Bewertung von Vitaminversorgungsmessgrößen. Niederkleen, Wissenschaftlicher Fachverlag Dr Fleck, 1993.
13 Rosenberg IH, Miller JW: Nutritional factors in physical and cognitive functions of elderly people. Am J Clin Nutr 1992;55:1237S–1243S.
14 Heseker H, Schneider R: Chronically increased vitamin intake and vitamin status of healthy men. Nutrition 1993;9:10–17.
15 Spector R: Megavitamin therapy and the central nervous system; in Briggs MH (ed): Vitamins in Human Biology and Medicine. Boca Raton, CRC Press, 1981, pp 137–156.
16 Kleijnen J, Knipschild P: Niacin and vitamin B_6 in mental functioning: A review of controlled trials in humans. Biol Psychiatry 1991;29:931–941.

17 Blom W, van den Berg GB, Huijmans JGM, Przyrembel H, Fernandes J, Scholte HR, Sanders-Woudstra JAR: Neurologic action of megadoses of vitamins. Bibl Nutr Dieta. Basel, Karger, 1986, No 38, pp 120–135.

18 Bonke D, Nickel B: Improvement of fine motoric movement control by elevated dosages of vitamin B_1, B_6, and B_{12} in target shooting; in Walter P, Stähelin H, Brubacher G (eds): Elevated Dosages of Vitamins. Toronto, Huber, 1989, pp 198–204.

Dr. H. Heseker, University GH Paderborn, Warburger Strasse 100,
D–33098 Paderborn (Germany)

Walter P (ed): The Scientific Basis for Vitamin Intake in Human Nutrition.
Bibl Nutr Dieta. Basel, Karger, 1995, No 52, pp 56–65

Folic Acid and the Prevention of Neural Tube Defects: The Need for Public Health Action

Nicholas J. Wald

Department of Environmental and Preventive Medicine,
Wolfson Institute of Preventive Medicine, The Medical College of
St. Bartholomew's Hospital, London, UK

Neural Tube Defects

Neural tube defects are ubiquitous; they have been found in all communities where they have been studied. Rates of neural tube defects in different communities range between about 0.5/1,000 births and about 3/1,000 – a 6-fold variation. Even a rate of 0.5/1,000 represents a rate that poses a serious medical problem. About half of all cases of neural tube defect have spina bifida alone, and half have anencephaly with or without spina bifida.

Primary Prevention of Neural Tube Defects

The strongest evidence that lack of dietary folic acid is a cause of neural tube defects comes from the combined results of four randomized prevention trials [Laurence et al., 1981; MRC Vitamin Study Research Group, 1991; Czeizel and Dudas, 1992; Kirke et al., 1992] (table 1). There were 40 neural tube defect pregnancies in these studies, and the rate of neural tube defects in the supplemented women was 76% lower than in the women receiving no supplements (relative risk 0.24; 95% confidence interval 0.11–0.52). The results show folic acid to be the protective agent – not other vitamins that are sometimes included in vitamin supplements. Non-randomized trials and observational studies yield consistent results. Protection is evident in women who have not had a pregnancy with a neural tube defect as well as in women who have.

Table 1. Folic acid and neural tube defects

Type of study	Number of studies	Dose of folic acid, mg	Number of neural tube defect pregnancies	Relative risk of a neural tube defect in the high folic acid category
Randomized trials	4	4 or 0.8	40	0.24 (0.11–0.52)
Non-randomized trials	2	0.4 or 5	31	0.12 (0.04–0.41)
Observational studies				
Including Mills et al.[1]	6	about 0.4	>1,000	0.47 (0.29–0.76)
Excluding Mills et al.[1]	5	about 0.4	>1,000	0.41 (0.27–0.61)

[1] The study of Mills et al. [1989] yielded a relative risk of 0.87 (0.73–1.02) using controls with congenital abnormalities other than a neural tube defect or 0.94 (0.80–1.10) using normal controls.

Folic acid supplementation, therefore, prevents about 3 out of 4 cases of neural tube defects. Since closure of the neural tube is completed by about 4 weeks after conception (about 6 weeks since the first day of the last menstrual period), it is important that the consumption of extra folic acid is started before a woman knows that she is pregnant. The observation that not all neural tube defects can be prevented by folic acid shows that these defects have more than one cause. The important conclusion is that *most* cases of neural tube defects are preventable by consuming sufficient folic acid immediately before pregnancy and in the early stages of pregnancy, although the precise mechanism of action is not known. From a public health perspective, a target intake of folic acid needs to be specified. On the basis of present evidence, an average extra intake of 0.4 mg folic acid is a reasonable target.

Three approaches can be adopted to increase folic acid consumption. Women planning a pregnancy can be given folic acid supplements; women can be encouraged to eat folate-rich foods, such as green leafy vegetables, asparagus and orange juice; and staple foods can be fortified with folic acid.

Supplementation

Although an extra 0.4 mg folic acid/day is a reasonable target, it may never be possible to determine the minimum fully effective supplement of folic acid. The results of several studies, including the Hungarian trial [Czeizel and

Dudas, 1992], indicate that supplements between 0.4 and 0.8 mg/day confer a substantial protective effect.

None of the randomized trials suggested any harm to the fetus or mother from folic acid supplementation, though the statistical power of the trials to assess this was limited. Two concerns have been raised: the exacerbation of the neuropathy of pernicious anaemia, and exacerbation of grand mal epilepsy in patients taking anticonvulsants.

The concern over pernicious anaemia is unjustified [Wald and Bower, 1995]. Folic acid does not cause pernicious anaemia. The fact that it may partially treat the anaemia but not the associated neuropathy is a reason to base the diagnosis on the B_{12} level, not the anaemia, and to treat pernicious anaemia with B_{12} and not folic acid. A dose of 4 mg/day can be given under medical supervision for women planning a pregnancy. A dose of 0.4 (or 0.8) mg/day is available without prescription for general use.

Women receiving anticonvulsants for epilepsy should be advised to take extra folic acid to reduce their risk of a neural tube defect pregnancy, but should be warned that there is a possibility that their epilepsy will be less well controlled. Blood levels of anticonvulsants should be monitored to ensure that satisfactory levels are maintained. The need to take folic acid is reinforced by the fact that certain anticonvulsants, notably valproic acid, can cause neural tube defects [Lindhout and Omtzigt, 1992].

The UK Government [Expert Advisory Group, 1992] has recommended that as well as eating more folate-rich foods, women who have previously had an affected child should take 4 or 5 mg/day folic acid supplements, and women in the general population should take a 0.4 mg/day supplement from the time they begin trying to conceive until the twelfth week of pregnancy. Such supplements are now readily available in chemist shops and health food shops. However, folic acid supplementation is not a practical method of reaching the majority of the population for two main reasons. Firstly, many pregnancies are unplanned, so advice to women planning a pregnancy will miss many preventable cases of neural tube defects. Secondly, neural tube defects are more common among the socio-economically disadvantaged groups in society, and these groups are less likely to be aware of the need to take folic acid.

Dietary Change

On average, people consume about 0.2 mg folate/day [Office of Population Censuses and Surveys, 1990]. Most studies have linked a preventive effect with daily intakes of three or more times this intake. To achieve such a change by

Table 2. Distribution of bread consumption among British women in 1987, and the maximum increase in intake of folic acid if bread were fortified with 0.3, 0.4 or 0.5 mg folic acid/100 g bread[1]

Bread g/day	Slices per day[2]	Cumulative percent of population	Maximum increase in intake of folic acid if bread is fortified with the specified amount of folic acid/100 g bread		
			0.3 mg	0.4 mg	0.5 mg
≤ 20	0.5	3.5	0.06	0.08	0.10
≤ 40	1	12.6	0.12	0.16	0.20
≤ 80	2	50.1	0.24	0.32	0.40
≤ 160	4	95.3	0.48	0.62	0.80
≤ 320	8	100.0	0.96	1.24	1.60

[1] From MAFF/DH Dietary Information Survey, 1987 [Dr. Buss, pers. commun. and Wald, 1993].
[2] Assuming 1 slice = 40 g.

changing diet is impractical without fortifying selected foods (e.g., bread and breakfast cereals) with folic acid.

Fortification

A barrier to introducing folic acid fortification of food is the concern that individuals will be unable to exercise choice over the composition of the food they eat. But the limitation to that choice is slight and the cost of such an absolute interpretation of individual autonomy great; families will unnecessarily continue to have babies with spina bifida. Most breakfast cereals and bread could be fortified, with a few brands still being available for those individuals who wanted unfortified foods.

A system of food fortification should be subject to guidelines, yet be flexible. In the case of folic acid fortification, the level can be selected to ensure that nearly everyone in the population consumes at least an extra 0.4 mg folic acid/day (i.e. 0.4 mg more than the natural median daily folate intake). Once the recommendation to fortify foods has been accepted by a national authority, there should be a standing committee to guide the food industry on the level of fortification.

Table 2 is an illustration of how, with information on the distribution of bread consumption in the population, the appropriate level of fortification can be estimated. In the example given, the fortification of bread with 0.5 mg/

100 g of bread would increase the median folic acid intake by about 0.4 mg/day.

Once set, the level of fortification should be reviewed periodically to determine whether the public health strategy had worked. A population survey showing that average serum levels had increased from 5 ng/ml (levels equivalent to an intake of about 0.2 mg/day) to 20 ng/ml (levels equivalent to women consuming an extra 0.3–0.4 mg/day of folic acid) [Schorah et al., 1983] would demonstrate that the strategy was working. Safety, as well as efficacy, would need to be reviewed. A monitoring unit could collect or assemble the necessary data on a systematic basis – for example by forming a national register of patients with B_{12}-deficient neuropathy.

The US Public Health Service recommended in 1992 that 'all women of childbearing age in the United States who are capable of becoming pregnant should consume 0.4 mg folic acid per day' [Morbidity and Mortality Weekly Report, 1922]. The UK Government [Expert Advisory Group, 1992] has recommended that the range of breads and breakfast cereals fortified with folic acid should be widened. They qualified this by adding that the present levels of folic acid fortification in breads and breakfast cereals should not be greatly exceeded. In the UK at present only a limited number of brands of bread are fortified with folic acid ('Mighty White', 'Champion', 'Co-op Soft Grain', 'Sainsbury's Soft Grain' and 'Tesco's healthy eating range'). Other brands and unbranded bread should be fortified so that most bread available contains extra folic acid. In the US an advisory committee to the Food and Drug Administration has also recommended fortification, and suggested flour as the food to be fortified. As flour is present in many foodstuffs, this strategy would also reach groups who rarely eat bread or breakfast cereal, but the level of fortification must be adequate. Figure 1 indicates how different levels of fortification of flour with folic acid will achieve the target intake.

The Notion of Prevention

Since only a proportion (albeit a large proportion) of neural tube defects can be prevented by folic acid supplementation, there has been a reluctance for public health authorities to say 'folic acid prevents neural tube defects', and alternative statements such as 'folic acid lowers the risk of neural tube defects' have been used, which dilutes the message. Few would object to saying that seat belts prevent fatal road traffic accident injuries, although, of course, they do not prevent all of them. To say that a nutrient prevents a medical disorder does not mean that it prevents all cases of that disorder.

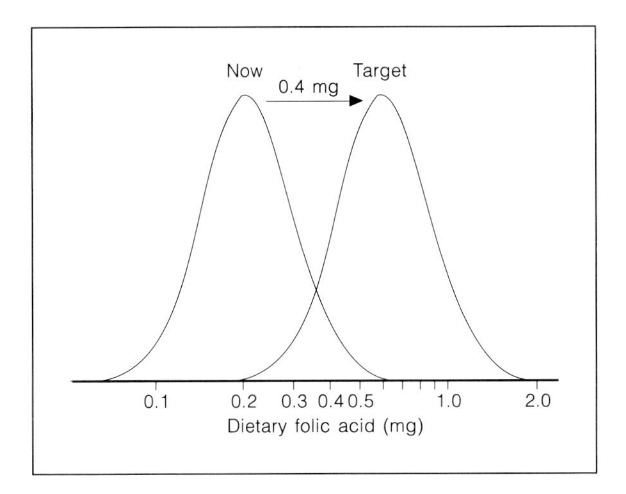

Fig. 1. Distribution of dietary folic acid per day.

Obstacles to Prevention

Denial of Personal Choice

'Fortification denies choice which is a human right'. This is too absolute a view. Many public health and safety measures are carried out on the basis of collective good. It may be expensive and impractical to offer a potential benefit allowing each individual an opportunity to make an explicit choice to accept or reject the benefit. The safety of buildings, the construction of vehicles and public roads, the treatment of waste, all involve collective decisions which, once in place, are there for all. A balance is needed between the collective benefit and individual choice. If, as well as the fortification of flour with folic acid, it is possible to have unfortified flour provided that this was specially labelled, choice would be preserved, but the burden of effort would fall on the consumer seeking the unfortified product.

Only a Small Minority Will Benefit

An argument against food fortification is that for every 10,000 people who may consume a fortified food, only one may benefit. The 9,999 other people have nothing to gain; half will be men, and of the women, many will not become pregnant. This is, however, a criticism that would apply to nearly all preventive measures and one that has been discussed by Rose [1981] and called, by him, the Prevention Paradox. For most public health strategies, each participating individual has a small chance of benefitting (e.g., dietary reduction of salt, stopping smoking, eating a low fat diet, traffic safety measures) and some may not benefit at all directly (e.g., boys vaccinated against rubella to prevent congenital rubella syndrome) but, on average, such collective action will have a large impact on the overall incidence of the diseases concerned. In the absence of being able to predict precisely who will benefit, it is not possible to restrict the preventive measures effectively. As long as the preventive measure, such as the fortification of food with folic acid, is affordable and acceptably safe, there is no reason to withold it from the general public, and every reason to introduce it so the benefits are realized.

Complete Safety Cannot Be Assured

There is no reason to believe that folic acid supplementation is harmful, but it is unreasonable to expect that any intervention will never have an adverse effect. The requirement that complete safety be assured would deny virtually all effective public health strategies. What is required is a balanced view with a quantitative estimate of proven benefit compared to plausible hazard, taking into account all relevant evidence. It is an issue that needs to be resolved with knowledge and pragmatism rather than by the inflexible application of absolute philosophical principles. It is important that regulatory bodies such as the US Food and Drug Administration do not show more concern over possible side effects that may accompany a proven method of prevention than the concern over the definite harm that will be done by not introducing it at all.

No Financial Incentive

A major obstacle to prevention arises, paradoxically, when the preventive measure is simple, cheap, and in the public domain, because no one has a financial incentive to introduce the measure. In such circumstances the state must play a role and encourage the introduction of the preventive measure – if necessary by legislation if a voluntary code of practice fails.

Restrictions on Use of Health Claims

Health claims for dietary supplements are not allowed in the UK and many other countries. Health claims should be allowed on food packages if they are true, do not mislead, and are based on sound scientific evidence. The claim, 'Folic acid prevents neural tube defects' is accurate and useful and should be allowed. In July 1994 the US Food and Drug Administration allowed certain health claims if they were soundly based, including one on folic acid and neural tube defects. It judges the evidence of health benefit according to the test of whether there is 'significant scientific agreement among qualified experts'.

Table 3. Folic acid fortification: present UK policy and proposed European Union policy

	UK now	European Union 'proposed'
Recommended daily allowance (RDA)	0.3 mg FA	0.14 mg FA
Claims per unit of food		
Standard	16.6% of RDA	15% of RDA
Strong	50% of RDA	30% of RDA
Unit of food consumption	serving per day	100 g
Examples:		
Breakfast cereal with strong claim	0.150 mg FA	0.013 mg FA
(1 serving = 30 g)	(0.3 mg × 50%)	(0.14 mg × 30% × 30/100)
		9% of present level
Bread with strong claim	0.150 mg FA	0.038 mg FA
(1 serving = 90 g)	(0.3 mg × 50%)	(0.014 × 30% × 90/100)
		25% of present level

FA = Folic acid.

Table 4. Folic acid fortification: present UK policy and desired policy

	UK now	Desired
Recommended daily allowance (RDA)	0.3 mg FA	0.6 mg FA
Claims per unit of food		
Standard	16.6% of RDA	33% of RDA
Strong	50% of RDA	67% of RDA
Unit of food consumption	serving per day	serving per day
Examples:		
Breakfast cereal with strong claim	0.150 mg FA	0.400 mg FA
(1 serving = 30 g)	(0.3 mg × 50%)	(0.6 mg × 67%)
		$2\frac{2}{3}$ of present level
Bread with strong claim	0.150 mg FA	0.400 mg FA
(1 serving = 90 g)	(0.3 mg × 50%)	(0.6 × 67%)
		$2\frac{2}{3}$ of present level

FA = Folic acid.

Lowering the Recommended Daily Allowance of Folic Acid

Under the UK food labelling regulations of 1984 a recommended daily allowance (RDA) of 0.3 mg folic acid is used. There are three changes that have been proposed by the Scientific Committee for Food: (1) lowering the RDA; (2) revision of the proportion of the RDA in a unit of food linked to a 'standard' and 'strong' claim on the packet, and (3) changing the unit of food per day (tables 3, 4). Each change would reduce the amount of folic acid fortification by about half. The three changes together would reduce the fortification to less than 10% for breakfast cereals and about 25% for bread.

In the interests of public health and safety this issue requires urgent consideration. The European Union needs to be advised on the issue with a view to increasing the labelling RDAs in the light of the new information on folic acid and neural tube defects, so that the labelling rules on claims and on the unit of daily food consumption can be revised, to encourage an average increase in folic acid intake of 0.4 mg folic acid/day. It is important that dietary RDAs should incorporate concepts of risk reduction for 'chronic' diseases such as neural tube defects where sufficient data for efficacy and safety exist.

Conclusion

Fortification of flour with folic acid is simple, economical and safe. Failure to take appropriate regulatory action will mean that tens of thousands of women in Europe will needlessly continue to have pregnancies with neural tube defects.

References

Czeizel AE, Dudas I: Prevention of the first occurrence of neural tube defects by periconceptional vitamin supplementation. N Engl J Med 1992;327:1832–1835.

Expert Advisory Group: Folic Acid and the Prevention of Neural Tube Defects. London, Department of Health/HMSO, 1992.

Kirke PN, Daly LE, Elwood JH: A randomised trial of low dose folic acid to prevent neural tube defects. Arch Dis Child 1992;67:1442–1446.

Laurence KM, James N, Miller MH, Tennant GB, Campbell H: Double-blind randomised controlled trial of folate treatment before conception to prevent recurrence of neural tube defects. Br Med J 1981;282:1509–1511.

Lindhaut D, Omtzigt JGC: Pregnancy and the risk of teratogenicity. Epilepsia 1992;33:S41–S48.

Mills JL, Rhoads GG, Simpson JL, Cunningham GC, Conley MR, Lassman MR, Walden ME, Depp RO, Hoffman HJ, and the National Institute of Child Health and Human Development Neural Tube Defect Study Group: The absence of a relation between the periconceptional use of vitamins and neural-tube defects. N Engl J Med 1989;321:430–435.

Morbidity and Mortality Weekly Report: Recommendations for the Use of Folic Acid to Reduce the Number of Cases of Spina Bifida and Other Neural Tube Defects. Atlanta, US Department of Health and Human Services, Sept 11, 1992, vol 41.
MRC Vitamin Study Research Group (prepared by N Wald with assistance from J Sneddon, J Densem, C Frost and R Stone): Prevention of neural tube defects: Results of the MRC Vitamin Study. Lancet 1991;338:132–137.
Office of Population Censuses and Surveys: The Dietary and Nutritional Survey of British Adults. London, HMSO, 1990.
Rose G: Strategy of prevention: Lessons from cardiovascular disease. Br Med J 1981;282:1847–1851.
Schorah CJ, Wild J, Hartley R, Sheppard S, Smithells RW: The effect of periconceptional supplementation on blood vitamin concentrations in women at recurrence risk for neural tube defect. Br J Nutr 1983;49:203–211.
Smithells RW, Sheppard S, Schorah CJ: Vitamin deficiencies and neural tube defects. Arch Dis Child 1976;51:944–949.
Wald N: Folic acid and the prevention of neural tube defects. Maternal nutrition and pregnancy outcome. Ann NY Acad Sci 1993;678:112–129.
Wald NJ, Bower C: Folic acid and the prevention of neural tube defects: A population strategy is needed. Brit Med J 1995;310:1019–1020.

Nicholas J. Wald, MBBS, DSc, Department of Environmental and Preventive Medicine, Wolfson Institute of Preventive Medicine, The Medical College of St. Bartholomew's Hospital, Charterhouse Square, London ECIM 6BQ (UK)

Walter P (ed): The Scientific Basis for Vitamin Intake in Human Nutrition.
Bibl Nutr Dieta. Basel, Karger, 1995, No 52, pp 66–74

Vitamins in the Maintenance of Optimum Immune Functions and in the Prevention of Phagocyte-Mediated Tissue Damage and Carcinogenesis

Ronald Anderson, Violet L. Van Antwerpen

Medical Research Council Unit for Inflammation and Immunity, Department of Immunology, Institute for Pathology, University of Pretoria, South Africa

Vitamins appear to interact beneficially with the human immune system in two distinct ways. Several vitamins (A, B_6 and D) appear to be primarily involved in promoting optimum immune functions, while others, most notably the antioxidant vitamins (C and E) and β-carotene, appear to have a primary containment function, enabling the immune system to function efficiently in the setting of minimal oxidant-inflicted tissue damage.

Vitamins and the Immune System

One of the most remarkable aspects of the human immune system is its resilience, which is characterized by the astonishing versatility, amplification potential and reserve capacity inherent in its cellular effector mechanisms. Immunodeficiency diseases such as chronic granulomatous disease (CGD) and AIDS highlight the magnitude of this reserve capacity. CGD is a congenital X-linked immunodeficiency in which phagocytes are unable to generate antimicrobial reactive oxidants, leading to severe, recurrent bacterial infections. Phagocytes from the mothers of these children have 50% of the normal oxidant-generating capacity, but unlike the children, the mothers are not prone to bacterial infections [1], demonstrating at least 50% reserve in the oxidant-mediated antimicrobial activity of these cells. Likewise, in HIV-positive individuals, opportunistic infections develop only when the numbers of circulating

$CD4^+$ T lymphocytes decline to 20–30% of normal levels [2]. Given the reserve capacity inherent in these cellular immune effector mechanisms, it is probable that only severe and prolonged vitamin deficiency states are likely to predispose to an increased frequency of infective episodes.

While deficiency of some vitamins may cause acquired immune dysfunction and increased incidence of infectious diseases, decreased intake of others may lead to overactivity of the cellular elements of the immune system, particularly phagocytes, leading to an increased threat of oxidant-mediated degenerative diseases and cancer. Given the magnitude and persistence of the public health problems presented by these diseases [3, 4], it is clear that the identification of discerning vitamin-based, public health immunopreventive strategies is a priority objective. Those vitamins for which the evidence of beneficial immunomodulatory effects is at least promising and at best compelling are reviewed here. Firstly, those vitamins which are involved in promoting optimum, protective host immune responses by nonoxidative mechanisms, namely vitamins A, B_6 and D, and secondly the antioxidant vitamins C and E and β-carotene which appear to prevent tissue damage mediated by immune system-derived oxidants. Vitamin B_{12} and folic acid, which primarily affect haematopoiesis, have not been included.

Nonantioxidative Vitamins and Host Defences

Of the nonantioxidative vitamins, A, B_6 and D appear to be intimately involved in maintaining efficient host defences.

Vitamin A

The evidence in favour of a critical role for vitamin A in maintaining host defences against microbial pathogens is particularly compelling. In developing countries, where vitamin A deficiency is common, this vitamin has been found to protect children against diarrhoeal disease, respiratory infections, severity of infections, complications of measles and mortality ascribed to different types of infection [reviewed in 5]. Two strategies have been used to prevent vitamin A deficiency in children. Firstly, regular supplementation in the form of capsules through a comprehensive or disease-targeted programme [5]. Capsules are administered every 4–6 months at doses ranging between 50 and 200,000 IU or 1.2×10^6 vitamin A equivalents of β-carotene [6] to rapidly replete and maintain blood and tissue levels. At these doses vitamin A and β-carotene appear to be equally effective [6]. Food fortification programmes based on the addition of vitamin A or β-carotene to regularly-consumed foodstuffs, as well as dietary diversification and nutrition education directed at consumption of vitamin A- and carotenoid-rich foods, offer an alternative strategy [5, 6]. These strategies are associated with decreased infectious morbidity and a 30–50% decrease in overall mortality [5]. In the therapeutic setting the WHO has advocated vitamin A supplements for children with severe measles, the recommended dose for children over age of 1 year being 200,000 IU daily for 2 consecutive days. This strategy is accompanied by significant decreases in both the frequency of infective complications and mortality [5].

Although the anti-infective properties of vitamin A are well recognized, a convincing mechanistic link between vitamin A deficiency and impaired cellular immunocompetence has yet to be demonstrated [7]. Maintenance of epithelial integrity and nonspecific mechanical barriers represents the most likely mechanism [7]. Vitamin A-mediated maintenance of epithelial integrity may promote optimum functions of ciliated epithelial cells and optimize leucocyte trafficking on and across the epithelium. To our knowledge there is no consistent evidence in support of impairment of either cellular immune effector mechanisms or synthesis and transport of secretory IgA during marginal vitamin A deficiency [7].

Correction of vitamin A deficiency in children appears to be the primary clinical application of supplementation with this vitamin [5]. Supplementation of vitamin A-sufficient individuals to achieve supraphysiological levels is apparently not associated with additional clinical benefits [7]. In this setting, β-carotene would appear to be anyway more appropriate, due to circumvention of vitamin A-induced toxicity and for the added potential benefit of the superior antioxidant properties of this agent.

Vitamin B_6

Severe experimentally-induced deficiency of vitamin B_6 in laboratory animals is primarily associated with disturbances in the maturation and functions of T lymphocytes [8–10]. In elderly humans, marginal deficiency of this vitamin is associated with decreased numbers and proliferative activities of circulating T lymphocytes, as well as interleukin-2 production by these cells [9, 10]. These acquired immunological abnormalities in humans are corrected by short-term (about 6 weeks) supplementation with 50 mg of vitamin B_6 daily [9, 10]. However, to our knowledge, there are no data demonstrating that long-term marginal deficiency of this vitamin predisposes to infection. Until such data are available, no recommendations regarding supplementation can be made solely on immunological grounds.

Vitamin D

The evidence implicating vitamin D in the maintenance of protective host immune responses is somewhat less compelling than that for vitamins A and

B_6. Nevertheless, potentially important evidence does exist in support of a role for this vitamin in augmenting host immunodefences against tuberculosis and cancer. The re-emergence of tuberculosis has recently been described as representing a global emergency [11]. In the case of tumour immunity, the apparent lack of tumour immunogenicity is perceived to be the major problem confronting cancer immunotherapy.

The antituberculosis activity of vitamin D was recognized in the 1940s, when it was used to treat skin TB [reviewed in 12]. More recently, an apparent association was identified linking hypovitaminosis D to an increased incidence of TB in recent Asian immigrants to the UK [12]. Hypovitaminosis D has therefore been suggested to be a predisposing factor for development of TB [12]. However, it is only recently that a possible immunological basis for this relationship has been identified. It has been demonstrated that human monocytes cultivated with the cytokine γ-interferon or bacterial endotoxin express 1-hydroxylase activity enabling these cells to generate $1,25(OH)_2D_3$ [12]. Exposure of monocytes to physiologically relevant concentrations of this agent in vitro promotes differentiation of these cells and acquisition of mycobacteriocidal activity [12]. This so-called $1,25(OH)_2D_3$-mediated autocrine activation of monocytes has been proposed to represent a potentially important mechanism of immune-mediated elimination of *Mycobacterium tuberculosis.*

Vitamin D has also been reported to enhance antitumour immunodefences, but these effects are achieved by different mechanisms involving modulation of the activity of immunosuppressive cells [reviewed in 13]. Several groups of researchers have reported that various types of cancer, including glioblastoma, lung, breast and hepatocellular carcinoma, produce the cytokine granulocyte/macrophage colony-stimulating factor [13]. This cytokine stimulates myelopoiesis leading to the production of a poorly-defined immune suppressor cell which originates in the bone marrow and which inactivates tumour-specific immune responses [13]. It has recently been reported however, that coincubation of tumour tissue with supraphysiological concentrations of $1,25(OH)_2D_3$ combined with γ-interferon in vitro, but not in the presence of the individual agents, inhibits the synthesis of granulocyte/macrophage colony-stimulating factor [13]. The consequence is restoration of antitumour immunity which is characterized by expansion and activation of tumour infiltrating $CD8^+$ T lymphocytes [13].

Although these apparent antituberculosis and antitumour immunomodulatory properties of vitamin D are interesting, no recommendations for supplementation can be made until additional data are forthcoming. In the case of tuberculosis the following must be established: (a) the possible predisposing role of hypovitaminosis D in the development of tuberculosis; (b) the effects of the disease per se, as well as those of chemotherapeutic agents, on vitamin D status, and (c) the safety of vitamin D in tuberculosis [12]. The proposed antitumour properties of this vitamin are based on in vitro data and are therefore preliminary. Moreover, the high concentrations of $1,25(OH)_2D_3$ which are required to achieve these effects indicate that this is a pharmacological rather than a nutritional issue, which will necessitate the use of analogues of $1,25(OH)_2D_3$ that do not cause hypercalcemia [13].

Antioxidant Vitamins and Host Defences

It is our belief that the primary interaction of the antioxidant vitamins C and E and β-carotene with the human immune system is one of containment of the wayward, pro-oxidative activities of phagocytes, thereby preventing oxidant-mediated tissue damage [14]. The immunostimulatory properties of these antioxidant nutrients are also well recognized and have been the subject of several recent reviews [15–17]. Severe experimentally-induced deficiency states in experimental animals are associated with decreased immune reactivity and increased susceptibility to experimentally-acquired infection, while supplementation in some cases leads to enhancement of immune function [15–17]. Experimentally-induced marginal vitamin C deficiency in humans is associated with acquired abnormalities of cell-mediated immunity [18]. However, while the effects of long-term marginal deficiency of the antioxidant vitamins and β-carotene on the frequency of infective episodes in humans is not known, it is becoming increasingly evident that marginal deficiencies of these nutrients may predispose to degenerative diseases and cancer. Enhancement of oxidation-sensitive immune functions is therefore likely to be a secondary function of these antioxidant nutrients, the primary function being one of containment of phagocytes.

The containment function of vitamins C and E and β-carotene is directed at the abundant, highly aggressive, oxidant-generating phagocytic cells of the immune system [1, 14]. The sheer numbers of these cells, as well as the undiscerning nature of their arsenal of toxic antimicrobial oxidants, constitutes an unremitting threat to other host cells and tissues. When designing this phagocyte-mediated host defence system, Nature obviously perceived that the critical survival advantages outweighed the constant threat of oxidant-mediated tissue injury and carcinogenesis, which could anyway be contained by innate antioxidant defences of which vitamins C and E and β-carotene are critical components [14]. Coexistence between the host and the phagocytic cells of his or her immune system is however fragile, and if disrupted, oxidant-mediated tissue injury may ensue. The importance of phagocytic cells in host defence, as well as their potential threat to other host cells and tissues, is emphasized by the astonishing tempo of production (1.5×10^9/kg b.w./day) of neutrophils, the predominant short-lived, circulating, professional phagocyte. Like the other cells of the immune system, these cells originate in the bone marrow with about 55–60% of marrow dedicated to their production [19]. Phagocytes possess a unique membrane-associated NADPH-oxidase which enables them to transform molecular oxygen into a series of potent oxygen-derived free radicals and reactive oxidants including superoxide, hydroxyl radicals, singlet oxygen, hydrogen peroxide and hypochlorous acid [reviewed in 19]. These oxidants are primarily antimicrobial and therefore protective, but these indiscriminate agents also endanger the host since they are immunosuppressive, proproteolytic, proadhesive, proatherogenic and carcinogenic [14, 19].

Several common, often avoidable, aspects of lifestyle may disrupt oxidant/antioxidant homeostasis. Poor dietary habits may compromise antioxidant defences through decreased intake of the antioxidant vitamins and β-carotene, while cigarette smoking, occupational and environmental atmospheric pollution and excessive exposure to ultraviolet radiation may cause a futile, potentially harmful increase in the numbers and pro-oxidative activities of circulating phagocytes [14, 20].

Antioxidant Nutrients, Phagocytes, Cardiovascular Disease and Cancer

There is now an abundance of evidence from epidemiological studies which shows that the circulating leucocyte, and the neutrophil count in particular, measured well before the development of manifest clinical disease is an *independent* predictor of several cardiovascular conditions including myocardial infarction, sudden cardiac death, all coronary heart disease combined, stroke and essential hypertension, as well as lung cancer incidence and mortality, possibly cancer at all sites and death from all causes [reviewed in 21–23]. For each decrease in circulating leucocyte of 1,000/ml, the risk of coronary heart disease death decreased by 14% [21], while the relative odds for a 2,000/ml difference in leucocyte count for development of lung cancer ranged from 1.20 to 1.58 in three different populations [22].

The circulatory system is particularly vulnerable to phagocyte-inflicted damage, since these cells are continually tumbling along, colliding with and adhering to vascular endothelium. These transient interactions with vascular endothelium, as well as transient intravascular aggregation of these cells, probably causes sustained generation and release of reactive oxidants leading to the proatherogenic, oxidative modification of low density lipoproteins.

The relationship between increased numbers of circulating leucocytes and future development of cancer may be attributable to the carcinogenic properties of phagocyte-derived reactive oxidants. Alternatively, the increased consumption of antioxidants due to elevated numbers of leucocytes may lead to secondary tissue depletion of these protective agents and increased vulnerability to endogenous, potentially carcinogenic reactive oxidants generated during cellular respiration.

Importantly, the antioxidant vitamins and β-carotene have been demonstrated to reduce cardiovascular mortality and incidence of cancer in numerous large epidemiological studies [24–27] and in some [28, 29], but not all [30] of the few intervention studies which have been completed. These aspects have been addressed in detail by others in this issue. The proposed mechanistic link between increased numbers of circulating phagocytes, increased requirement for antioxidant nutrients and development of oxidant-mediated degenerative diseases and cancer is strengthened by our recent observations that plasma levels of vitamin C [31] and β-carotene [pers. unpubl. observations] are significantly and inversely correlated with the numbers of circulating leucocytes and neutrophils. In the case of vitamin E however, positive correlations between this antioxidant and the number of circulating leucocytes and neutrophils have been observed [31] indicating (a) that vitamin E may function as a mobilizable antioxidant and (b) that depletion of vitamin E in tissues might not be reflected by plasma levels of the vitamin.

Recommended Daily Intakes of Vitamins C and E and β-Carotene

We have used data from a recently completed, unpublished study on cigarette smoking to calculate the optimum daily intake of the antioxidant nutrients based on the numbers and pro-oxidative activities of circulating leucocytes. Cigarette smoking has a profound irritant effect on phagocytes and is characterized by a chronic, futile inflammatory response in the lungs, as well as increased numbers of circulating leucocytes, particularly neutrophils, increased oxidant release by these cells and decreased levels of vitamin C and β-carotene in the circulation and of vitamin E in the lungs [reviewed in 31]. The average increase in the circulating leucocyte count of smokers in our study was 2×10^6/ml (6.2×10^6/ml and 8.2×10^6/ml for 85 male nonsmokers and 100 age- and sex-matched cigarette smokers respectively). The average increase in the smoking-related, phagocyte-mediated oxidant burden was calculated to be 71% (due to increased numbers and pro-oxidative activities of circulating leucocytes). Based on this increase in the oxidant burden and on RDAs for vitamin C, vitamin E and β-carotene (no RDA for this agent at present) of

60, 15 and 6 mg/day respectively, the estimated increases in the daily intakes (above the RDA levels) for each increment of 2×10^6/ml in the circulating leucocyte count above 6.2×10^6/ml are 40 mg vitamin C, 10 mg vitamin E and 4 mg β-carotene. Given that the highest circulating leucocyte count observed in our study group of apparently healthy smokers was 13×10^6/ml, the total estimated daily intakes of the antioxidant nutrients required to protect all smokers are 200, 50 and 20 mg of vitamin C, vitamin E and β-carotene respectively. There are, however, several flaws in the calculation such as: (a) it is based on RDAs which may or may not be optimal; (b) in vitro activation of phagocytes may not reflect the in vivo situation, and (c) it may not be necessary to increase the intake of all three antioxidants. Nevertheless, these values are in good agreement with estimates from metabolic [32], experimental [33] and epidemiological studies [24, 34].

In conclusion, harmonious coexistence with the phagocytic cells of the immune system is probably dependent on optimum availability of vitamins C and E and β-carotene. The circulating leucocyte count may therefore be a primary determinant of optimum intake of these antioxidant nutrients.

References

1 Gallin JI: Phagocytic cells: Disorders of function; in Gallin JI, Goldstein IM, Snyderman R (eds): Inflammation – Basic Principles and Clinical Correlates. New York, Raven Press, 1988, pp 493–511.

2 Ameisen JC, Capron A: Cell dysfunction and depletion in AIDS: The programmed cell death hypothesis. Immunol Today 1991;12:102–105.

3 Berkelman RL, Hughes JM: The conquest of infectious diseases: Who are we kidding? Ann Intern Med 1993;119:426–428.

4 Beardsley T: A war not won. Sci Am 1994;Pt 1:118–126.

5 Labadarios D: Vitamin A – Time for action. South Afr Med J 1994;84:1–2.

6 Carlier C, Coste J, Etchepare M, Périquet B, Amédée-Manesme O: A randomised controlled trial to test equivalence between retinyl palmitate and β-carotene for vitamin A deficiency. BMJ 1993; 307:1106–1110.

7 Ross AC: Vitamin A status: Relationship to immunity and the antibody response. Proc Soc Exp Biol Med 1992;200:303–320.

8 Meydani SN: Micronutrients and immune function in the elderly. Ann NY Acad Sci 1990;587: 196–207.

9 Miller LT, Kerkvliet NI: Effect of vitamin B_6 on immunocompetence in the elderly. Ann NY Acad Sci 1990;587:49–54.

10 Meydani SN, Ribaya-Mercado JD, Russell RM, Sahyoun N, Morrow FD, Gershoff SN: The effect of vitamin B_6 on the immune response of healthy elderly. Ann NY Acad Sci 1990;587:303–306.

11 Godlee F: Tuberculosis – 'A global emergency'. BMJ (South Afr ed) 1993;i:694.

12 Rook GW: The role of vitamin D in tuberculosis. Am Rev Resp Dis 1988;138:768–770

13 Young MRI, Halpin J, Wang J, Wright MA, Matthews J, Schmidt Pak A: 1α,25-Dihydroxyvitamin D_3 plus γ-interferon blocks lung tumor production of granulocyte-macrophage colony-stimulating factor and induction of immunosuppressor cells. Cancer Res 1993;53:6006–6010.

14 Anderson R: Assessment of the roles of vitamin C, vitamin E and β-carotene in the modulation of oxidant stress mediated by cigarette smoke-activated phagocytes. Am J Clin Nutr 1991;53(suppl): 358–361.
15 Anderson R, Smit MJ, Jooné GK, Van Staden AM: Vitamin C and cellular immune functions: Protection against hypochlorous acid-mediated inactivation of glyceraldehyde-3-phosphate dehydrogenase and ATP generation in human leukocytes as a possible mechanism of ascorbate-mediated immunostimulation. Ann NY Acad Sci 1990;587:34–48.
16 Kelleher J: Vitamin E and the immune response. Proc Nutr Soc 1991;50:245–249.
17 Bendich A: β-Carotene and the immune response. Proc Nutr Soc 1991;50:263–274.
18 Jacob RA, Kelley DS, Pianalto FS, Swendseid ME, Hennig SM, Zhang JZ, Ames BW, Fraga CJ, Peters JH: Immunocompetence and oxidant defense during ascorbate depletion of healthy men. Am J Clin Nutr 1991;54(suppl):1302–1309.
19 Anderson R: The activated neutrophil-formidable forces unleashed. South Afr Med J 1995; in press.
20 Savage JE, Theron AJ, Anderson R: Activation of neutrophil membrane-associated oxidative metabolism by ultraviolet radiation. J Invest Dermatol 1993;101:532–536.
21 Grimm RH Jr, Neaton JD, Ludwig W: Prognostic importance of the white blood cell count for coronary cancer, and all-cause mortality. JAMA 1985;254:1932–1937.
22 Phillips AN, Neaton JD, Cook DG, Grimm RH, Shaper AG: The leukocyte count and risk of lung cancer. Cancer 1992;69:680–684.
23 Friedman GD, Fireman BH: The leukocyte count and cancer mortality. Am J Epidemiol 1991;133: 376–380.
24 Gey KF, Puska P, Jordan P, Moser UK: Inverse correlation between plasma vitamin E and mortality from ischemic heart disease in cross-cultural epidemiology. Am J Clin Nutr 1991;53(suppl):326–334.
25 Trout DL: Vitamin C and cardiovascular risk factors. Am J Clin Nutr 1991;53(suppl):322–325.
26 Riemersma RA, Wood DA, MacIntyre CCA, Elton RA, Gey KF, Oliver MF: Risk of angina pectoris and plasma concentrations of vitamins A, C and E and carotene. Lancet 1991;337:1–5.
27 Block G: Micronutrients and cancer: Time for action? J Natl Cancer Inst 1993;85:846–848.
28 Blot WJ, Li J-Y, Taylor PR, et al: Nutrition intervention trials in Linxian, China: Supplementation with specific vitamin/mineral combinations, cancer incidence, and disease-specific mortality in the general population. J Natl Cancer Inst 1993;85:1483–1492.
29 Li J-Y, Taylor PR, Li B, et al: Nutrition intervention trials in Linxian, China: Multiple vitamin/ mineral supplementation, cancer incidence, and disease-specific mortality among adults with esophageal dysplasia. J Natl Cancer Inst 1993;85:1492–1498.
30 The Alpha-Tocopherol and Beta-Carotene Cancer Prevention Study Group: The effect of vitamin E and beta-carotene on the incidence of lung cancer and other cancers in male smokers. N Engl J Med 1994;330:1029–1035.
31 Van Antwerpen L, Theron AJ, Myer MS, Richards GA, Wolmarans L, Booysen U, Van der Merwe CA, Sluis-Cremer GK, Anderson R: Cigarette smoke-mediated oxidant stress, phagocytes, vitamin C, vitamin E, and tissue injury. Ann NY Acad Sci 1993;686:53–65.
32 Kallner AB, Hartmann D, Hornig DH: On the requirements of ascorbic acid in man: Steady-state turnover and body pool in smokers. Am J Clin Nutr 1981;34:1347–1355.
33 Frei B, England L, Ames BN: Ascorbic acid is an outstanding anti-oxidant in human blood plasma. Proc Natl Acad Sci USA 1989;86:6377–6381.
34 Schectman GR: Estimating ascorbic acid requirements for cigarette smokers. Ann NY Acad Sci 1993;686:335–346.

Dr. R. Anderson, Institute for Pathology, PO Box 2034, Pretoria 0001 (South Africa)

Walter P (ed): The Scientific Basis for Vitamin Intake in Human Nutrition.
Bibl Nutr Dieta. Basel, Karger, 1995, No 52, pp 75–91

Cardiovascular Disease and Vitamins

Concurrent Correction of 'Suboptimal' Plasma Antioxidant Levels May, as Important Part of 'Optimal' Nutrition, Help to Prevent Early Stages of Cardiovascular Disease and Cancer, Respectively

K. Fred Gey

Vitamin Unit, Institute of Biochemistry and Molecular Biology,
University of Berne, Switzerland

Basic Epidemiological Data

There is overwhelming observational evidence that diets rich in antioxidant micronutrients (e.g. the vegetarian-type or Mediterranean diets) are associated with a lower risk of premature death from cardiovascular disease (CVD) and cancer respectively [1, 2]. Within the large series of nutritionally active plant constituents, antioxidants became highly intriguing when it was accepted that reactive oxygen species are able to damage all cellular compounds and might modulate gene expression. An imbalance between pro-oxidative factors and the antioxidative defense system ('oxidative stress') has been implicated particularly with early stages of atherogenesis and cancerogenesis [3–11]. The working hypothesis is that an 'optimal' antioxidant status is a prerequisite of 'optimum health' as defined by WHO.

The assay of plasma levels of vitamins A, C, E and carotene in complementary epidemiological studies (cross-cultural comparisons, cases of early untreated angina pectoris, and the individual risk of mortality in prospective studies) is a logical and promising approach to define the adequacy of the antioxidant status in vivo regarding the risk of cancer and CVD. The hitherto accumulated, still incomplete but consistent data permit the tentative assumption [1, 5, 9–11] of *threshold plasma levels associated with minimal relative risk (desirable 'optimum' levels) of premature death by CVD and cancer*, i.e. >50

(-60) μmol/l vitamin C, >30 μmol/l lipid-standardized vitamin E (α-tocopherol/cholesterol ratio ≥ 5.1–5.2), >2.2 μmol/l vitamin A, and >0.4 (-0.5) μmol/l β-carotene or >0.4–0.5 μmol/l α-plus β-carotene. Plasma levels being >25–35% lower predict an approximately 2-fold risk of cancer and CVD respectively but they are still higher than levels of classical overt vitamin deficiency [9,11].

Regarding cancer mortality β-carotene, as representative carotenoid, has most frequently revealed an inverse correlation (particularly regarding lung cancer) whereas associations of other antioxidants have been inconsistent [5–7], e.g. of vitamin E [8]. Inconsistencies might, at least in part, be related to the missing variation of a given plasma antioxidant and/or to a cohort's uniform antioxidant status above the threshold of risk [1]. The most regular inverse correlations for coronary heart disease (CHD) has been found for lipid-standardized vitamin E, but vitamin C and carotenoids such as β-carotene can be involved independently of a fairly high vitamin E status. In fact, the relative risk of CVD can be significantly increased at a suboptimal level of any single antioxidant, and thus the combination of low levels of two antioxidants increases the risk additively to overmultiplicatively [1, 5, 10, 11]. All this suggests for the human multiple synergistic interactions of essential antioxidants as observed in vitro and in animals [1]. Therefore, the major causes of premature death may be related to a 'suboptimal' overall antioxidant potential rather than the 'suboptimal' status of any particular antioxidant [11]. A 'suboptimal' status of many antioxidants tested thus far, but particularly of vitamin E, may be of primary importance in Northern Europe whereas vitamin C and carotenoids can be relevant risk factors also in Central Europe [11].

For the multifactorial CVD, suboptimal levels of antioxidants can be stronger and more frequent risk predictors of CHD than classical risk factors, such as hyperlipidemia and hypertension [1, 5, 9–11]. Potential risks by suboptimal antioxidant levels have a great public health impact since 'optimal' plasma levels (lacking the association with material risk) are achievable by adequate common diets without supplements, e.g. in Central and Southern Europe [1, 9–11].

Intervention Data

Epidemiological studies on steady-state differences of diet-derived antioxidant micronutrients in relation to disease can obviously be confounded in many ways. In principal, any causal relationship remains finally to be proven by properly controlled experiments on specific supplements in randomized subjects of high risk and initially inadequately poor antioxidant status. But supplementation of antioxidants late in life, e.g. simply a few years prior to the common aging-related rise of mortality, evaluates only therapeutic potentials in the last stages of CVD and cancer respectively, and this design has even pitfalls (see below). Logically, proper studies on primary prevention, the principal goal of public health, require intervention for several decades of life which meet, however, problems of feasibility, compliance, funding, etc. Thus, for the time being, observational data on long-term intake and plasma levels of micronutrients still remain the fundamental source of information. They are most conclusive when regular self-supplementation corrects both a prevalent 'suboptimal' antioxidant status and the increased relative risk of CVD and cancer respectively. This review is going to focus on conditions of such studies.

First US National Health and Nutrition Examination Survey (NHANES I)

In representatives of the general middle-aged US population, as surveyed in NHANES I (mean age 53 years of age, including 24–34% smokers), the voluntary habitual self-supplementation of vitamin C (mean 130 mg daily in comparison to the clearly insufficient mean intake of 22 mg daily in supplement nonusers) reduced the mortality from all CVD, i.e. by about 42% (95% confidence interval 22–59%) in males, and by 25% (confidence interval 1–45%) in females, but only when vitamin C was consumed within multivitamin preparations [12]. This indicates that in self-supplementing US adults (many of which may actually have practiced a generally health-oriented lifestyle) the health benefits of vitamin C depended on the concurrent adequacy of other micronutrients. Indeed, a substantial percentage of US adults has an insufficient intake of other vitamins as well, e.g. of vitamin E, B_6, folate and retinol [2, 11, 13], and this deficit might have been rectified by the multivitamin preparation. The latter may have, according to the RDA 1968, contained fair amounts of both vitamin A (1.5 mg) and vitamin E (30 IU = 20 mg) as well as the daily requirements of folate and other B vitamins. Vitamin C-containing multivitamin supplements of NHANES I lowered cancer mortality by only 22% in males, and 15% in females (both without statistical significance), but these effects may have contributed to the reduction of total mortality by 35% in males (confidence interval 20–48%), and in both sexes by 33% (confidence interval 11–33%) [12]. A specific role of vitamin C remains conceivable in spite of the fact that its health benefit clearly depended on the concurrent 'optimization' of other components of multivitamin supplements [12]. Thus, the overall recording of multivitamin supplementation without specific consideration of vitamin C or of other constituents lacked any correlation to total mortality in NHANES I [14]. Since multivitamin supplementation is frequently part of a health-oriented lifestyle, a major influence of the latter does not seem to be likely either. Health benefits by 130 mg vitamin C daily in the general population fit the above suggested 'optimum' plasma level of vitamin C as deduced from classical epidemiology. From data of NHANES II and other studies it may be assumed that the regular intake of 130 mg vitamin C achieves on one hand in nonsmokers more or less tissue saturation with vitamin C which is reached at renal threshold levels at approximately 65–75 (–85) μmol/l. On the other hand, in smokers (whose requirement is typically increased) 130 mg daily may yield plasma levels at the lower end of the desirable range ≥ 50 μmol/l [11, 15, 16]. Thus, the prospective NHANES I evaluation of mortality by CVD and cancer [12] supports the above mentioned observational data which suggest that an 'optimal' vitamin C status is a (one of many) prerequisite(s) for preventing premature death in middle-aged males.

US Health Professionals' Study and US Nurses' Health Study

Very special strata of the US population such as male middle-aged health professionals (mean age 54 years, including only 10% smokers and 22–31% regular aspirin users, with consumption of an 'ideal' low fat diet of about 30% lipid calories) revealed specific benefits of the habitual 'optimization' of vitamin E and β-carotene respectively [17]. In men the self-supplementation of vitamin E (mean ≥ 67 mg daily, mostly as part of multivitamin preparations) decreased the relative risk of coronary events by 46% (confidence interval 22–67%; $p < 0.01$ for linear trend in multivariate analysis) in comparison to supplement nonusers with an especially low mean intake (< 4.4 mg, i.e. less than half the present US RDA of 10 mg, and about half of 8 mg consumed by the general US population [11, 18]). In health professionals, plasma levels with the desirable α-tocopherol/cholesterol ratio of 5.1 were achieved with vitamin E supplements [19], in contrast to supplement nonusers whose ratio was presumably < 2.8, i.e. most likely in a range predicting a severalfold increased risk of CHD [11]. Corresponding protective effects of vitamin E supplements were found in US nurses (approximately 50 years old, 23–28% smokers; fat calories 15%; 50–57% regular aspirin users) [20]. In male health professionals, β-carotene supplements (mean 9 mg daily in comparison to 3 mg in supplement nonusers) failed to affect the risk of nonsmokers but in smokers the β-carotene supplements reduced the coronary risk by 70% (confidence interval 18–89%; $p < 0.02$ for linear trend in multivariate analysis) [17]. Supplements of the above order of β-carotene achieved 'optimum' plasma level (mean 0.46 μmol/l) [18]. The different effect of β-carotene supplements in nonsmoking and smoking health professionals may be related to the fact that even in nonsupplementing nonsmokers the plasma level of β-carotene was above the proposed 'optimum' threshold of > 0.40 μmol/l whereas smokers revealed only 0.30 μmol/l [21], i.e. a level predicting an increased CHD risk [11]. Desirable 'optimum' plasma levels may be achieved in nonsmokers by a daily intake of approximately 2–3 mg but smokers may require approximately $\geq$9 mg [11, 16]. Smokers tend to be typically low in plasma β-carotene, probably in part due to an increased requirement [11, 16, 21]. Health professionals were unsuited to test specific benefits of vitamin C [17] since even in supplement nonusers the vitamin C intake was fairly high (160 mg) [22] which may have yielded at least 'optimal' if not saturation plasma levels in both nonsmokers and smokers [15, 16]. Supplements of B vitamins lacked also any significant correlation to CHD events [17] as could be expected from the fact that their supply was close to the RDAs even in nonsupplementing health professionals (again in contrast to the general US population), and supplementing males reached at least twice the RDA [22]. Thus, the fact that supplements other than vitamin E and β-carotene lacked any significant health benefits in multivariate analysis [17] does not exclude that the full preventive potentials of vitamin E and β-carotene (in analogy to those of vitamin C in NHANES I) required synergistic interactions with 'optimal' levels of other micronutrients. In conclusion, this special US cohort shows a specific requirement of vitamin E 'optimization' for coronary health, and an additional need of rectifying of β-carotene in smokers, both independently from a concurrently fair supply of other micronutrients. But nevertheless the latter could still be a condition of the beneficial effects of vitamin E and β-carotene respectively regarding cardiovascular health.

Breast Cancer Evaluation in US Nurses

Whereas reports on the relative cancer risk of health professionals are still missing, the relative prospective risk of breast cancer was reported for US nurses [23]. Increase of vitamin C intake above the minimum intake of 93 mg daily failed to affect the cancer risk as suggested by the fact that this intake probably yielded 'optimum' plasma levels at least in nonsmokers [15]. Habitual voluntary supplementation of vitamin E ($>$183 mg daily in comparison to $<$4 mg) tended slightly to decrease the multivariate breast cancer risk (up to 20%, but with $p = 0.07$ only). As with coronary events [20] the intake of other vitamins in multivitamin preparations lacked any correlation to breast cancer with the exception of vitamin A which at $>$3 mg daily resulted in a reduction up to 47% (confidence interval 23–46%; $p = 0.02$ for linear trend in multivariate analysis) of the relative breast cancer risk due to low intake of preformed vitamin A from food ($<$1.77 mg) [22]. Vitamin A might be particularly important in two respects, i.e. on one hand for sex hormone-related tumors, e.g. of prostate and breast, and on the other hand regarding its interaction with β-carotene [24]. In conclusion, in US nurses, whose nutritional status was probably comparable to that of US health professionals, the fairly specific correction of a previously 'suboptimal' status of vitamins A and E rectified also the increased relative breast cancer risk but again the clinical effect might have depended on 'optimal' levels of other micronutrients.

Retrospective Case-Control Study on Lung Cancer in Non-Smokers/Ex-Smokers in New York State

In this large population-based survey [25] vitamin E-containing supplements were associated with risk reduction by 45% (confidence interval 15–65%). This seems plausible with regard to the fact that in the general US population as screened in NHANES II, the vitamin E intake from diet including fortified food is as low as 8 mg daily [18] which can only yield plasma levels predicting a substantially increased risk [11]. The reduction of the relative risk of lung cancer by vitamin E supplements was confounded by higher consumption of raw fruits, greens (and dietary β-carotene respectively) as well as milk and cheese. In consequence, it remains again to be elucidated whether the benefits of rectifying vitamin E required the concurrent 'optimization' of other nutrients.

Study on the Risk of Oral and Pharyngeal Cancer in Four Areas of the US

A population-based prospective follow-up of voluntarily multivitamin-supplementing subjects (both sexes; mean age ~59 years, 52% smokers) revealed that the long-term regular intake of vitamins A, B, C, and E within multivitamin preparations significantly reduced the relative risk of oral and pharyngeal cancer adjusted for race, smoking and alcohol [26]. After adjustment for 'ever regularly used' vitamin E in doses of ≥67 mg daily, the significance for other vitamins got lost. Vitamin E intake was related to a 50% reduction of relative risk (confidence interval 40–60%; $p < 0.001$ for trend with dose and time). It remains again to be clarified whether concurrent optimization of other nutrients by the multivitamin preparations was a condition of the protective potential of vitamin E.

Iowa Women's Health Study on Colon Cancer

In this prospective study on colon cancer (mean age 62 years) habitual voluntarily multivitamin supplements were associated with a significant reduction of the age-adjusted relative risk in age-adjusted univariate analysis [27]. Thereby supplemental vitamin A (≥1.5 mg daily) reduced by 53% (95% confidence interval 21–73%; trend 0.002), vitamin C (≥60 mg) by 33% (10–80%; trend 0.01), and vitamin E (≥20 mg) by 56% (29–72%; trend 0.0002). In multivariate analysis protective effects remained statistically significant for vitamin A (43% reduced relative risk; trend $p < 0.04$) and for vitamin E (≥24 mg; mean 39 mg; 50% reduced risk; trend $p < 0.01$). Nevertheless, because of the concurrent consumption of these vitamins (correlated with $r \sim 0.4–0.5$) it remains again to be considered whether the reduction of colon cancers by vitamins A and/or E demanded a concurrently optimal status of other antioxidants such as vitamin C. The age- and vitamin A-adjusted cancer-protective effect of vitamin E was strongly age-dependent: supplemental vitamin E reduced the risk of women 55–59 years of age by 84% (95% confidence interval 30–96%), in women 60–64 years old by 63% but only without statistical significance, and at >65 years of age not at all [27]. This variable interaction between vitamin E and age clearly suggests that vitamin E can only counteract earlier stages of carcinogenesis but may hardly attenuate the age-related progression of clonically established tumors.

Multicomponent Intervention Trials in Linxian, China

In the middle-aged general population (both sexes, mean 52 years; 30% smokers) of Linxian, a rural area of poor nutritional status of Northern China [28], the randomized supplementation of a combination of β-carotene, vitamin E and selenium (15 mg, 30 mg, and 50 μg respectively) resulted in a 10% reduction of the mortality from strokes, i.e. the second most important cause of death (CHD being of marginal importance in the Far East). The raw data published thus far indicate that the reduction reached the order of >20% when the actual co-administration of either vitamin C (120 mg) plus molybdenum or of riboflavin (3.2 mg) plus niacin [28] was taken into account. In a parallel randomized trial testing in subjects with prevalent dysplasia of the esophagus/cardia [29], the benefit of fixed combination of all vitamins and trace elements (antioxidant doses as above, other micronutrients mostly in the order of two RDAs) was increased further, i.e. stroke mortality in men decreased by 45% ($p<0.05$) [29, 30]. This clearly suggests that the concurrent optimization of all essential micronutrients substantially increases the chance of attenuating CVD. In the general population the combination of β-carotene, vitamin E and selenium reduced also the esophagus/stomach cancer rate by 21% ($p<0.05$), and total cancer mortality by 13% ($p=0.05$). The latter was more effectively reduced by special combinations, i.e. by 24% when β-carotene, vitamin E and selenium were supplied together with riboflavin plus niacin (3.2 and 40 mg respectively), and by 29% when β-carotene, vitamin E and selenium were supplemented together with vitamin A and zinc (5,000 IU and 23 mg) [28]. Thus cancer risk may also have a greater chance to be diminished by multifactorial optimization than by supplementing a few selected antioxidants. In patients with established, fairly well advanced and almost irreversible precancerous lesions (dysplasia of the esophagus/cardia) even the above mentioned multivitamin/multimineral preparation failed to reduce cancer mortality. The missing protection of patients with established dysplasia [29] indicates that preventive potentials of the tested micronutrients are restricted to very early stages of the disease [28, 30]. This assumption is on one hand in line with consistent observations in non-Caucasians and Caucasians that specific supplements of β-carotene and/or vitamin A [31–33] as well as by vitamin E [34, 35] suppress very early precancerous stages such leukoplakia/metaplasia and micronuclei (e.g. by chewing mutagenic betel nuts and/or tobacco and by cigarette smoking respectively). On the other hand, no hard experimental data have ever been suggested that antioxidants (which scavenge or quench reactive oxygen species as implicated with initiation and promotion of cancerogenesis) can also ameliorate irreversible tumor progression of later stages [3, 4, 6, 36]. In the subjects of the Linxian trials the antioxidant plasma levels rose from an initially suboptimal range to that fairly above the risk threshold, e.g. vitamin C from <14 to 47 μmol/l, β-carotene from <0.21 to 1.59 μmol/l [28, 29]. Thus, the Chinese data fit again the above suggested 'optimal' plasma ranges.

Finnish Alpha-Tocopherol, Beta-Carotene Cancer Prevention Study in Chronic Heavy Smokers

In the placebo group of this randomized intervention trial (chronic male smokers 50–69 years of age at entry) subjects in the lowest quartile of serum vitamin E or β-carotene had a considerably higher incidence of lung cancer than subjects in the highest quartile of serum antioxidants [37]. This result fully concurs with abundant observational evidence [2, 5, 8, 38] that the long-term or the lifetime intake of either substance in the regular diet reduces the relative risk of cancer [39–41]. In contrast, single or combined supplements of an exceptionally well absorbable β-carotene preparation (20 mg daily singly or in combination with vitamin E) for 6.1 years lacked any substantial benefits and possibly even accelerated the progression of lung cancer by 18% and of CHD by 12% [37]. Vitamin E supplements (50 IU all *rac*-α-tocopheryl aceate ≈34 mg *RRR*-α-tocopherol) reduced prostate cancer mortality by 34% ($p<0.05$) but failed to reduce the overall and lung cancer mortality as well as CVD mortality [37]. Since the vitamin E supplements clearly corrected the initially poor plasma vitamin E levels (from 26.7 μmol/l with an α-tocopherol/cholesterol ratio of 4.3 at entry to 'optimal' levels of 40 μmol/l with an α-tocopherol/cholesterol ratio of 6.5 [37, 42]) this relatively short-termed improvement of the vitamin E status probably came 'too late' for the chronic smokers. After approximately 30 years of heavy smoking the majority of study subjects may have fairly progressed, complicated blood vessel lesions with minimal chance towards regression as well as cancer cell clones in the lung [36]. The assumption that vitamin E came 'too late' to stop the progression of lung cancer, is strongly supported by the fact that vitamin E supplements significantly reduced the incidence of prostate cancers [37]. Thus, prostate cancer in contrast to the lung cancer was hardly regularly established at baseline since this sex hormone-related cancer is strongly promoted during the 'male menopause' which probably emerged during the test period in subjects with initial mean age of 57. Accordingly, Swiss males of the same age category developed an up to 18-fold rise of the relative risk of prostate cancer (adjusted for age and other variables) if their plasma vitamin E was at baseline the lowest quartile [43]. Aside from coming 'too late', the lacking protective effects of single supplements of vitamin E and/or β-carotene in chronic Finnish smokers may also involve 'too little' consideration of the underlying nutritional status of smokers. As generally known, smokers are notoriously poor not only in the antioxidants tested but also in many other vitamins, e.g. vitamin A, C, folate and some B vitamins [11]. For example, the calculated daily intake of 97 mg vitamin C in Finnish smokers [42] may have yielded only 'suboptimal' plasma levels of 37–47 μmol/l [15, 16]. It has to be recalled that vitamin C, the first line of antioxidative defense, is also required to regenerate vitamin E and to protect vital oxygen-sensitive compounds such as glutathione (the principal endogenous intracellular antioxidant) and folate (important for several metabolic pathways and cellular differentiation). In any case, the intervention protocol (copying designs for cancer therapy) proved to be insufficient to test the potentials of single antioxidants for primary prevention of CVD and common cancers respectively in elderly chronic Finnish smokers.

Secondary Cancer Intervention Trials in the US

The above conclusion is supported on one hand by the failure of secondary prevention by β-carotene of non-melanoma skin cancer (in patients <85 years old with about half <65 years of age, randomized after surgical removal of the primary tumor and supplemented for approximately 10 years) [44]. On the other hand, several reports consistently stated missing effects of antioxidants on the recurrence of colorectal adenomas. This was true for supplements of vitamins C plus E in young patients of familial polyposis (35 years of age, followed for 4 years after removal of the entire colon) whereas vitamins C plus E in combination with fibers clearly tended to reduce the recurrence of adenomas [45]. In the elderly with isolated polyps (2-year follow-ups after removal of visible adenomas, mean age of patients 58 years, 80% smokers) vitamin C plus E reduced the recurrence rate by 14% only without statistical significance [46]. β-Carotene either singly or combined with vitamins C plus E (for 4 years after polypectomy in patients mostly 51–70 years of age, the majority at 61–70) did also not reduce the recurrence rate [47]. In contrast, supplements of combined vitamins A, C and E given for 6 months after removal of isolated visible adenomas (patients 40–80 years of age, mean 64 years) reduced significantly precancerous cell proliferation in normal-appearing rectal mucosa [48]. In conclusion, the combined vitamins A, C and E may reduce very early precancerous cell abnormalities in the colon/rectum even in old age but apparently vitamins fail to inhibit the proliferation of previously established adenoma clones escaping deduction by initial colonoscopy. Clearly, the clinical outcoming of adenomas (similar to that of lung tumors in the above mentioned Finnish intervention trial) cannot be reduced by a relatively short supplementation of a few selected antioxidants whereas long-term (or ideally lifelong) consumption of diets high in vegetables and fruits by providing antioxidant together with a variety of nutritionally active compounds including fibers may have better chances. As mentioned above, multivitamin preparations relatively rich in vitamins A and E reduced the risk of forthcoming colon cancer [27]. All this points again to the principle of preventing multifactorial diseases by multifactorial 'optimization', i.e. by eliminating all complementary pathogenic mechanisms as far as possible.

Where To Go From Here?

In case of 'significant scientific agreement among qualified experts' the FDA can permit health claims that diets or supplements rich in antioxidants such as vitamins C, E and β-carotene reduce the risk of cancer and CVD [39, 49]. But the above survey reveals manifold difficulties in a scientific interpretation and integration of results from the 'gold standard' of intervention, i.e. the double-blind, population-based randomized intervention shortly before increasing mortality (as actually designed for drug therapy). The results from the Finnish intervention trial [37, 42] are not easy to reconcile with those from interventions in China [28–30] and the extensive observational evidence with and without self-supplementation which indicates that the risk of cancer and CVD is significantly lower at habitual and probably long-term consumption of antioxidants provided by diets rich in fruits and vegetables and from appropriate multivitamin preparations respectively. Both contain, of course, a series of nutrients and nonnutrient factors other than antioxidants, the identification of which may at present be the most exciting scientific task in this field. Clearly, future population-based randomized intervention trials, which can conclusively test primary prevention over extended periods, will be a big undertaking that can hardly become routine. Therefore, new ways will have to fill the gap of knowledge. Two main avenues look most promising, i.e.:

(1) Many more observational studies for:

(a) Final figures on steady-state plasma levels of hitherto studied micronutrients. If plasma levels of antioxidant micronutrients were reliable indicators of adequacy, a consensus on 'optimal' levels will be highly desirable (conceivable after completion of current studies, e.g. the Vitamin Substudy of the WHO/MONICA Project [9], the prospective Basle Study [5, 24, 43] and the PRIME Study [50]). Generally accepted data are needed for conclusive risk assessments in individuals and communities as well as for intake recommendations, based on pending consent on the response of plasma levels to regular dietary supply.

(b) New data on the relative importance and adequacy of presumably interacting 'orphan' nutrients, e.g. carotenoids other than β-carotene (e.g. lutein, β-cryptoxanthin, lycopene [51]), common plant phenols (e.g. quercetin [52, 53] and red wine bioflavonoids [54]), folate and other B vitamins (modulating many vitally important pathways, e.g. of homocysteine metabolism [55, 56]), minerals (potassium, magnesium, calcium, selenium and other trace metals), the monounsaturated oleic acid and special PUFAs, special leguminous proteins, unrefined carbohydrates, as well as nonnutrients such as potentially carcinostatic sulfur compounds (of garlic, cabbage, broccoli, etc.), and fibers of different texture.

(2) Improvement of existing and development of new functional test ('biomarkers') on biologically important cellular effects of nutrients which can be studied ex vivo and may be crucial for early steps of the pathogenesis of CVD and cancer respectively.

Most intriguing may be the study of:

(a) Entire native plasma/serum, e.g. regarding oxidizability, self-aggregability of LDL in serum, and its complex formation with proteoglycans typical for aged and/or arteriosclerotic blood vessels, etc.

(b) Plasma–cell interactions, e.g. between autologous serum and isolated thrombocytes and monocytes regarding cell adhesivity, transformation of monocyte-derived macrophages into foam cells, cytokine secretion of monocytes as well as of T and B lymphycytes, etc.

(c) Cell–cell interactions in autologous serum, e.g. between T cells and monocytes or between monocyte/macrophages and cultured human endothelium, as well as phagocytosis by macrophages of standard tumor cells.

(d) In vivo changes of precancerous lesions such as proliferation rate [48], leukoplakia/metaplasia and micronuclei [31–33].

(e) Indices of the body's overall reaction, e.g. regarding immune responses or the excretion of oxidized DNA bases as a result of common radical damage, e.g. by cigarette smoking.

Suitable functional tests applied to a relative small number of healthy volunteers and/or patients should permit testing of diets and of intriguing constituents including their interdependencies and interactions. Short-termed studies on selective depletion and repletion singly and in combinations are particularly appealing. Short experiments could also reveal the effect of any given nutrient singly and in combination within different diets, e.g. in 'American' versus 'Mediterranean' diets in genetically comparable test subjects. Of course, suitable 'biomarkers' may also allow an easy differentiation of constituents of interesting foods, e.g. of garlic, onions, fennel. Functional tests if combined with the assay of plasma levels offer also tools to study homeostatic or genetic factors for blood levels, the importance of which is presently almost unknown. Any detailed and hard evidence from such complementary functional test could become powerful catalysts for a (or a few) still desirable but technically difficult and costly long-term randomized intervention trial(s) to be especially designed for primary prevention. Extended observational data together with mechanistic short-term functional tests ('biomarkers') may reveal the relative importance of nutritionally active factors responsible for 'optimum' health. If preventive properties can finally be proven by one (to two) conclusive intervention trial(s) the entire evidence is likely to provide significant scientific agreement among qualified experts.

Safety and Recommendable Intake

The 'FDA continues to believe that the majority of vitamin and mineral supplements used today . . . do not raise safety concerns' [39], and this may also be true for the above mentioned supplements which revealed health-protecting properties in the US [17, 20, 23, 25–27] and in China [28–30] and for which no adverse effects were reported. For vitamins C and E the risks of health hazard by high doses seem practically to be negligible. Thus, (healthy) kidneys excrete any excess of vitamin C, and the intestinal absorption rate of vitamin E decreases exponentially at increasing intake. For β-carotene, however, it remains still to be explored whether the increase of cancer mortality in notorious Finnish smokers aged 50–69 years in response to 20 mg daily of a special highly absorbable β-carotene preparation is not a chance finding in spite of its statistical significance [37]. In fact, for other β-carotene supplementation trials no adverse affects have been reported thus far. But the latter employed less readily absorbable β-carotene preparations yielding mean plasma levels ≤ 3 μmol/l β-carotene [17, 28, 29, 44], in contrast to 5.6 μmol/l β-carotene in Finnish smokers [37]. It is widely known that any antioxidant, if no longer balanced within the antioxidant cascade, can reveal pro-oxidative properties, as least in vitro. Furthermore, rapidly proliferating tissues, including nonmalignant ones, tend to grade up their antioxidant status and might thus theoretically concentrate β-carotene. It must also be remembered that the excess of special liposoluble vitamins with extremely high biological potency and hormone-like actions, such as vitamins A and D, can provoke adverse effects. Regarding all presently open questions it seems prudent to avoid megadoses of any antioxidant and particularly plasma levels >3 μmol/l β-carotene. Megadoses of antioxidant micronutrients are also not justified by the available observational data. Thus, the latter indicate that antioxidant micronutrients may help to maintain health if their 'suboptimal' status is prevented or timely corrected but not from high intake per se.

The present although still incomplete studies on the dose response of plasma levels to oral antioxidant intake suggests that the majority of healthy middle-aged subjects of westernized societies can achieve an 'optimal' plasma status by intake orders being close to or only moderately above the present RDAs [11], i.e. by (a) approximately 60–80 mg vitamin C in nonsmokers, and approximately 125–130 mg in smokers; (b) presumably around 20 (15–30) mg vitamin E in Europe, but ≥67 mg in male Americans (differences presumably at least in part due [3] to vitamin E sources with an unfavorable low α-tocopherol/PUFA ratio and poor to negative 'net vitamin E' [57] respectively), and (c) 2–3 mg β-carotene in nonsmokers, and ≥9 mg in male smokers.

If according to WHO/FAO 'recommended intakes' should be 'amounts considered sufficient for the maintenance of (optimum) health for nearly all people', the same amounts seem to be appropriate. Thus, in middle-aged subjects their daily intake as voluntary or randomized supplements significantly reduced the relative risk of CVD and cancer respectively [17, 19, 20, 23, 26–30]. Recommendations for vitamin E should replace absolute figures by 'net vitamin E' and/or a dietary α-tocopherol (mg)/PUFA (g) ratio of 0.6 [57, 58].

Conclusions

Epidemiological surveys provided abundant evidence that under steady-state conditions diets rich in antioxidants (from vegetables/fruits and suitable vegetable oils) reduce the relative risk of premature death from CVD and cancer. Material relative risks seem to disappear at 'optimal' antioxidant plasma levels in the order of ≥50 μmol/l vitamin C, ≥30 μmol/l lipid-standardized vitamin E (α-tocopherol/cholesterol ratio ≥5.1–5.2), ≥2.2 μmol/l vitamin A, and ≥0.4 μmol/l β-carotene or ≥0.4–0.5 μmol/l α-plus β-carotene. Levels 25–35% below these thresholds predict an at least 2-fold higher risk. 'Suboptimal' levels of any single antioxidant may increase the relative risk independently. Accordingly, 'suboptimal' levels of several antioxidants predict a further increase of risk.

Data on habitual voluntary multivitamin supplements providing an adequate supply of either vitamins A, C or E, and of β-carotene in smokers, indicates that steady-state 'optimization' reduces more or less regularly the relative risk of CVD and cancer respectively. Simple counting of multivitamins regardless of their composition did not reveal any risk reduction. The antioxidant-related health benefits seem to depend on an adequacy of all antioxidants, and possibly of nonantioxidant nutrients as well. Thereby, an overall 'optimal' antioxidant defense system may be more important than excess of any particular 'magic bullet' antioxidant. Although antioxidants may represent a crucially important fraction within a health-maintaining diet, any nonantioxidant co-nutrients remain to be identified which could condition the health benefits of antioxidants.

In randomized antioxidant intervention trials during 5–6 years in middle-aged to elderly subjects in China and Finland, only earlier stages of CVD and cancer respectively were prevented by rectifying previously poor levels. Correspondingly, the incidence of prostate cancer (developing mostly not until the male menopause) was reduced by correction of a previously poor vitamin E status in Finland. In contrast, irreversible precancerous lesions (such as esophageal dysplasia), clonically established common cancers (highly probable for the lung of elderly heavy smokers) as well as (presumably advanced, complicated) vascular lesions of chronic smokers did not respond favorably. This strengthens public health's demands to start primary prevention as early in life as possible. As far as compared in China the efficacy of primary prevention was highest when antioxidant combinations were complemented by other vitamins and nonantioxidant micronutrients respectively. This concurs with the assumption that 'optimum health' (as defined by WHO) requires the adequate balance of all nutritionally active compounds rather than the selective improvements of special nutrients. This fits the widely accepted view that CVD and cancer are multifactorial diseases and therefore demand multifactorial risk factor intervention. Intriguing co-nutrients of antioxidants in health-promoting fruits and vegetables are phenols/bioflavonoids/anthocyanins, carotenoids other than β-carotene, folate, riboflavin, oleic acid, potentially carcinostatic compounds in garlic, broccoli, etc.

References

1 Gey KF: Vitamin E and other essential antioxidants regarding coronary heart disease: Risk assessment studies. Epidemiological basis of the antioxidant hypothesis of cardiovascular disease; in Packer L, Fuchs J (eds): Vitamin E: Biochemistry and Clinical Applications. New York, Dekker, 1993, pp 589–633.
2 Block G: The data support a role of antioxidants in reducing cancer risk. Nutr Rev 1992;50:207–213.

3 Ames BN, Shigenaga MK: Oxidants are a major contributor to cancer and ageing; in Halliwell B, Aruoma OI (eds): DNA and Free Radicals. Chichester, Ellis Horwood, 1993, pp 1–15.
4 Guyton KZ, Kensler TW: Oxidative mechanisms in carcinogenesis. Br Med Bull 1993;49:523–544.
5 Gey KF: Prospects for the prevention of free radical disease, regarding cancer and cardiovascular disease. Br Med Bull 1993;49:679–699.
6 Van Poppel G: Carotenoids and cancer: An update with emphasis on human intervention studies. Eur J Cancer 1993;29A:1335–1344.
7 Schorah CJ: Micronutrients, antioxidants and risk of cancer. Bibl Nutr Dieta. Basel, Karger, 1995, No 52, pp 92–107.
8 Knekt P: Epidemiology of vitamin E: Evidence for anticancer effects in humans; in Packer L, Fuchs J (eds): Vitamin E: Biochemistry and Clinical Applications. New York, Dekker, 1993, pp 513–527.
9 Gey KF, Moser UK, Jordan P, Stähelin HB Eichholzer M, Lüdin E: Increased risk of cardiovascular disease at suboptimal plasma levels of essential antioxidants: An epidemiological up-date with special attention of carotene and vitamin C. Am J Clin Nutr 1993;57(suppl):787S–797S.
10 Gey KF: Optimum plasma levels of antioxidant micronutrients. Ten years of antioxidant hypothesis on arteriosclerosis. Bibl Nutr Dieta. Basel, Karger, 1994, No 51, pp 84–99.
11 Gey KF: Ten-year retrospective on the antioxidant hypothesis of arteriosclerosis: Threshold plasma levels of antioxidant micronutrients related to minimum cardiovascular risk. J Nutr Biochem 1995; 6:206–236.
12 Enstrom JE, Kanim LE, Klein MA: Vitamin C intake and mortality among a sample of the United States population. Epidemiology 1992;3:194–202.
13 Surgeon General, US Department of Health and Human Services, Public Health Service: The Surgeon General's report on nutrition and health. Washington, Department of Health and Human Services, 1988.
14 Kim I, Williamson DF, Byers T, Koplan JP: Vitamin and mineral supplements use and mortality in a US cohort. Am J Public Health 1993;83:546–550.
15 Smith JL, Hodges RE: Serum vitamin C in relation to dietary and supplemental vitamin C in smokers and nonsmokers. Ann NY Acad Sci 1987;498:144–152.
16 Bolton-Smith C, Casey CE, Gey KF, Smith WCS, Tunstall-Pedoe H: Antioxidant vitamin intakes assessed using food frequency questionnaire: Correlation with biochemical status in smokers and non-smokers. Br J Nutr 1991;65:337–346.
17 Rimm EB, Stampfer MJ, Ascherio A, Giovannucci E, Colditz GA, Willett WC: Vitamin E consumption and the risk of coronary heart disease in men. N Engl J Med 1993;328:1450–1456.
18 Block G, Sinha R, Gridley G: Collection of dietary-supplement data and implications for analysis. Am J Clin Nutr 1994;59(suppl):232S–239S.
19 Ascherio A, Stampfer MJ, Colditz GA, Rimm EB, Litin L, Willett WC: Correlations of vitamin A and E intakes with the plasma concentrations of carotenoids and tocopherols among American men and women. J Nutr 1992;122:1792–1801.
20 Stampfer MJ, Hennekens CH, Manson JAE, Colditz GA, Rosner B, Willett WC: Vitamin E consumption and the risk of coronary heart disease in women. N Engl J Med 1993;328:1444–1449.
21 Rimm E, Colditz G: Smoking, alcohol, and plasma levels of carotene and vitamin E. Ann NY Acad Sci 1993;686:323–334.
22 Rimm E, Giovannucci E, Stampfer MJ, Colditz GA, Litin LB, Willett WC: Reproducibility and validity of an expanded self-administered semiquantitative food frequency questionnaire among male health professionals. Am J Epidemiol 1992;135:1114–1126.
23 Hunter DJ, Manson JE, Colditz GA, Stampfer MJ, Rosner B, Hennekens CH, Speizer FE, Willett WC: A prospective study of the intake of vitamins C, E, and A and the risk of breast cancer. N Engl J Med 1993;329:234–240.
24 Gey KF, Stähelin HB, Eichholzer M, Lüdin E: Prediction of increased cancer risk in humans by interacting suboptimal plasma levels of retinol and carotene; in Livrea MA, Packer L (eds): Retinoids. New Trends in Research and Clinical Application. Basle, Birkhäuser, 1994, pp 137–163.
25 Mayne ST, Janerich DT, Greenwald P, Chorost S, Tucci C, Zaman MB, Melamed MR, Kiely M, McKneally MF: Dietary beta-carotene and lung cancer risk in US nonsmokers. J Natl Cancer Inst 1994;86:33–38.

26 Gridley G, McLaughlin JK, Block G, Gluch M, Fraumeni JF Jr: Vitamin supplement use and reduced cancer risk of oral and pharyngeal cancer. Am J Epidemiol 1992;135:1083–1092.
27 Bostick RM, Potter JD, McKenzie DR, Sellers TA, Kushi LH, Steinmetz KA, Folsom AR: Reduced risk of colon cancer with high intake of vitamin E: The Iowa Women's Health Study. Cancer Res 1993;53:4230–4237.
28 Blot WJ, Li JY, Taylor PR, Guo W, Dawsey S, Wang GQ, Yang CS, Zheng SF, Gail M, Li GY, Yu Y, Liu B-q, Tangrea J, Sun Y-h, Liu F, Fraumeni JF Jr, Zhang YH, Li B: Nutrition intervention trials in Linxian, China: Supplementation with specific vitamin/mineral combinations, cancer incidence, and disease-specific mortality in the general population. J Natl Cancer Inst 1993;85:1483–1492.
29 Li JY, Taylor PR, Li B, Dawsey S, Wang GQ, Ershow AG, Guo W, Liu SF, Yang CS, Shem Q, Wang W, Mark SD, Zuo XN, Greenwald P, Wu YP, Blot WJ: Nutrition intervention trials in Linxian, China: Multiple vitamin/mineral supplementation, cancer incidence, and disease-specific mortality among adults with esophageal dysplasia. J Natl Cancer Inst 1993;85:492–1498.
30 Blot WJ, Taylor PR, Li B, et al: Nutrition intervention trials in Linxian, China. Am J Clin Nutr 1995; in print.
31 Stich H, Mathew B, Sankaranarayanan R, Nair KK: Remission of precancerous lesions in the oral cavity of tobacco chewers and maintenance of the protective effects of β-carotene or vitamin A. Am J Clin Nutr 1991;53:298S–304S.
32 Garewal H, Meyskens F, Friedman S, Alberts D, Ramsey L: Oral cancer prevention: The case for carotenoids and anti-oxidant nutrients. Prevent Med 1993;22:701–711.
33 Van Poppel G, Kok F, Hermus RJ: Beta-carotene supplementation in smokers reduces the frequency of micronuclei in sputum. Br J Cancer 1992;66:1164–1168.
34 Benner SE, Winn RJ, Lippmann SM, Poland J, Hansen KS, Luna MA, Hong WK: Regression of oral leukoplakia with α-tocopherol: A community clinical oncology program prevention study. J Natl Cancer Inst 1993;85:44–47.
35 Benner SE, Wargovich MJ, Lippmann SM, Fisher R, Velasco M, Winn RJ, Hong WK: Reduction in oral mucosa micronuclei frequency following alpha-tocopherol treatment of oral leukoplakia. Cancer Epidemiol Biomarkers Prev 1994;3:73–76.
36 Prior WA: Letter to the Editor. N Engl J Med 1994;331:612.
37 Heinonen OP, Albanes D for The Alpha-Tocopherol, Beta-Carotene Cancer Prevention Study Group: The effect of vitamin E and beta-carotene on the incidence of lung cancer and other cancers in male smokers. N Engl J Med 1994;330:1029–1035.
38 Block G, Patterson B, Subar A: Fruit, vegetables, and cancer prevention: A review of the epidemiological evidence. Nutr Cancer 1992;18:1–29.
39 Food and Drug Administration of the US: Antioxidants and cancer study. FAD Talk Paper 1994; T94-20:1–4.
40 Kritchevsky D: Letter to the Editor. N Engl J Med 1994;331:611–612.
41 Heinonen OP, Huttunen JK, Albanes D, Taylor PR for The Alpha-Tocopherol, Beta-Carotene Cancer Prevention Study Group: The author's reply to letters to the Editor. N Engl J Med 1994; 331:613.
42 Heinonen OP, Albanes D for The Alpha-Tocopherol, Beta-Carotene Cancer Prevention Study: Design, methods, participant characteristics, and compliance. Ann Epidemiol 1994;4:1–10.
43 Eichholzer M, Stähelin HB, Gey KF, Lüdin E, Bernasconi F: Prediction of cancer risk by suboptimal plasma levels of interacting antioxidants: 17-year follow-up of the Basel Prospective Study. 1994, submitted.
44 Greenberg ER, Baron JA, Stukel TA, et al and the Skin Cancer Prevention Group: A clinical trial of beta-carotene to prevent basal cell and squamous cell cancers of the skin. N Engl J Med 1990; 323:789–795.
45 DeCosse JJ, Miller HH, Lesser ML: Effect of wheat fiber and vitamins C and E on rectal polyps in patients with familial adenomatous polyposis. J Natl Cancer Inst 1989;81:1290–1297.
46 McKeown-Eyssen G, Holloway C, Jazmaji V, Bright-See E, Dion P, Bruce WR: A randomized trial in the prevention of recurrence of colorectal polyps. Cancer Res 1988;48:4701–4705.
47 Greenberg ER, Baron JA, Tosteson TD, et al and the Polyp Cancer Prevention Group: A clinical trial of antioxidant vitamins prevent colorectal carcinoma. N Engl J Med 1994;331:141–147.

48 Paganelli GM, Biasco G, Brandi G, Santucci R, Gizzi G, Villani V, Cianci M, Miglioni M, Barbara L: Effect of vitamin A, C and E supplementation in patients with colorectal adenomas. J Natl Cancer Inst 1992;84:47–51.
49 Department of Health and Human Services, Food and Drug Administration: Food Labelling: Health Claims and Label Statements; Antioxidant Vitamins and Cancer – Final Rule. Federal Register 1993;58:2622–2660.
50 Evans A: The PRIME Study. 1993
51 Ziegler RG, Subar AM, Craft NE, Ursin G, Patterson BH, Graubard BI: Does β-carotene explain why reduced cancer risk is associated with vegetable and fruit intake? Cancer Res 1992;52(suppl): 2060s–2066s.
52 Hertog MGL, Hollman PCH, Katan MB, Kromhout D: Intake of potentially anticarcinogenic flavonoids and their determinants in adults in the Netherlands. Nutr Cancer 1993;20:21–29.
53 Hertog MGL, Feskens EJM, Hollman PCH, Katan MB, Kromhout D: Dietary antioxidant flavonoids and risk of coronary heart disease: The Zutphen Elderly Study. Lancet 1993;342:1007–1011.
54 Maxwell S, Cruickshank A, Thorpe G: Red wine and antioxidant activity in serum. Lancet 1994; 344:193–194.
55 Ubbink JH, Vermaak WJH, Van Der Meerwe A, Becker PJ: Vitamin B_{12}, vitamin B_6, and folate nutritional status in men with hyperhomocysteinemia. Am J Clin Nutr 1993;57:47–53.
56 Ueland M, Refsum H, Stabler SP, Malinow MR, Andersson A, Allen RH: Total homocysteine in plasma or serum: Methods and clinical applications. Clin Chem 1993;39:1764–1779.
57 Bässler KH: On the problematic nature of vitamin E requirements: Net vitamin E. Z Ernährungswiss 1991;30:174–180.
58 Gey KF: Extra vitamin E beyond PUFA-dependent vitamin E requirement is supplied by olive oil and sunflower oil but not by soybean oil and other oils with insufficient α-tocopherol/PUFA ratio. Int J Vitamin Nutr Res 1995;65:61–64.

Prof. Dr. K.F. Gey, Vitamin Unit, Department of Biochemistry and Molecular Biology, University of Berne, POB, CH–3000 Berne (Switzerland)

Walter P (ed): The Scientific Basis for Vitamin Intake in Human Nutrition.
Bibl Nutr Dieta. Basel, Karger, 1995, No 52, pp 92–107

Micronutrients, Antioxidants and Risk of Cancer

C.J. Schorah

Division of Clinical Sciences, School of Medicine, The Old Medical School, University of Leeds, UK

An assessment of the scientific basis for vitamin intake in the prevention and treatment of cancer is a formidable task because the reviewer is faced with an enormous number of diverse reports. Fortunately, recent findings have provided the foundation for a framework which could explain how a number of apparently chemically diverse micronutrients have a common role in the prevention, and possibly the treatment of cancer. It is this framework, and the evidence for it, that will be examined in this review. It concerns primarily vitamins A, C and E, the carotenoids, and possibly folic acid. The common role for these micronutrients is their potential to maintain the structure and function of DNA, often, though not exclusively, through antioxidant mechanisms. Evidence for this role, including the chemical and metabolic perspectives and the findings of observational and intervention studies in man, will be examined.

Prevention of Carcinogenesis

Our understanding of the carcinogenic process has developed considerably during the last 10–15 years [1–3]. Essentially the mechanism is considered to be a multistep process occurring over several decades. Figure 1 indicates the major stages in this process, with each stage possibly requiring more than one change in cell metabolism. Initiation is believed to be a primary mutation which possibly activates an oncogene or inhibits an antioncogene, or both. The change to DNA structure could include base deletion or adduct formation and strand breaks or cross-linking [4, 5]. Whilst such damage to DNA is

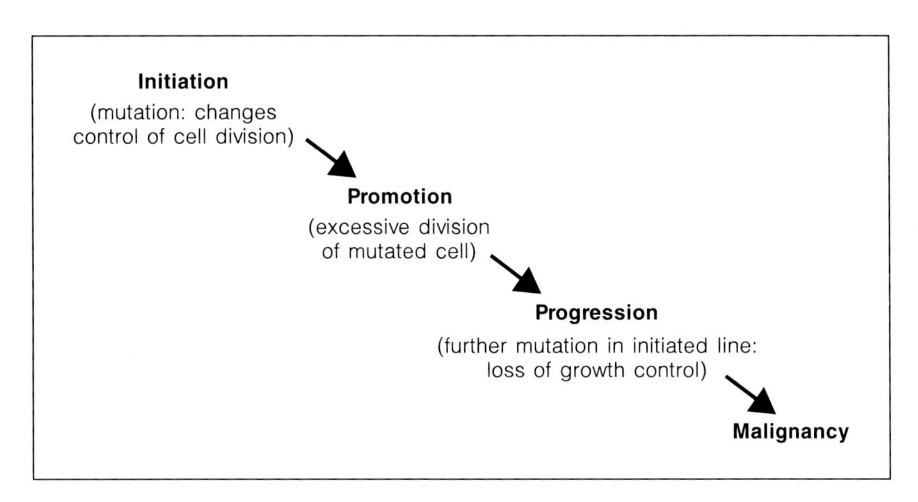

Fig. 1. Major stages in the carcinogenic process. Each stage may require more than one step and is capable of being activated by reactive species, including pro-oxidants and free radicals (see text for details).

probably not uncommon [1, 6] there are very effective processes within the cell which are able to repair the damage and prevent and change leading to a permanent alteration in the DNA code [5, 6]. However, occasionally these repair mechanisms fail, and replication of the fault results in a permanent mutation. When this mutation affects oncogene or antioncogene control it can result in inappropriate cell growth in the presence of a promoter. Promoters are themselves usually under metabolic control and so the inappropriate growth of these mutated cells is limited and they struggle to survive, particularly in the presence of constraints imposed by surrounding normal cells and immune mechanisms. However, if metabolic processes favour activation of promoters, the mutated cells can divide rapidly forming small clusters of cells or 'benign microtumours'. If this process, known as promotion, continues, encouraged by conditions which stimulate metabolism and are associated with malignancy, sufficient numbers of cells are produced in these clusters to allow a further chance mutation to occur in one of these cells, which eliminates the need for the presence of the promoter for cell growth. This is progression and as this cell can now divide in the absence of the promoter, growth of this cell type is uncontrolled, and they are on the threshold of malignancy. Full malignant potential may require further modification, such as the ability of the mutated cells to overcome growth restriction imposed by surrounding tissue.

Any single step in the process of carcinogenesis is probably fairly likely to happen, but the body has protective mechanisms against all these changes.

This defence prevents all but a few cells in a lifetime reaching the stage where malignant transformation is close to completion.

We are as yet unclear what chemicals and metabolites control the steps in this model of carcinogenesis, but recent research has indicated that the redox potential of the cell may be central in maintaining a balance between carcinogenesis and its prevention. Pro-oxidant states and reactive species (such as oxygen-centred free radicals like superoxide and hydroxyl) seem to encourage the processes of tumourigenesis outlined in figure 1, whereas antioxidants, which can scavenge or quench reactive species, have the opposite effect.

The evidence for this link with reactive species has been considered in a number of recent reviews [2–12]. Situations where reactive species can be generated in excess, such as chronic infection, smoking and exposure to radiation, considerably increase the risk of malignancy. Reactive species themselves, plus chemical systems which lead to their generation, are able to cause structural changes in DNA causing cells in culture and in animal studies to mutate. Several carcinogens are either reactive species or are able to generate these species and some potential carcinogens become much more active in the presence of free radicals. There is, therefore, good evidence for the involvement of reactive species and free radicals in cell mutation, a step believed to control both the initiation and progression steps of the carcinogenic process. Finally, it would seem that the process of promotion, which may not involve mutagenesis, is normally controlled by the balance between pro-oxidants and reducting agents in the cell [3, 13, 14].

If excess reactive species and pro-oxidant activity encourages carcinogenesis, antioxidants and scavengers of reactive species should be protective. There is increasing evidence that this is the case [3, 5, 11]. Many intracellular antioxidants are proteins, such as superoxide dismutase, glutathione peroxidase, and catalase [7]. Supporting the role of these enzymes within the cell, but also acting in the extracellular compartment, are metabolites such as uric acid and bilirubin and protein thiol groups. However, micronutrient antioxidants such as vitamins A, C, E and the carotenoids also have a major extracellular antioxidant function and can also be effective in cell membranes and within the cell itself. In addition, retinoids, such as vitamin A, may have protective activities other than antioxidant function in maintaining cell integrity and differentiation [15–17]. Here we must also include folic acid which, although not an antioxidant, may contribute to inhibiting the process outlined in figure 1 by methylating DNA and protecting it from reactive species attack [18]. As all these substances are part of a normal diet, and as they are (with the exception of some retinoids) relatively nontoxic, their concentrations in biological fluids and cells are relatively easily and safely manipulated by changes

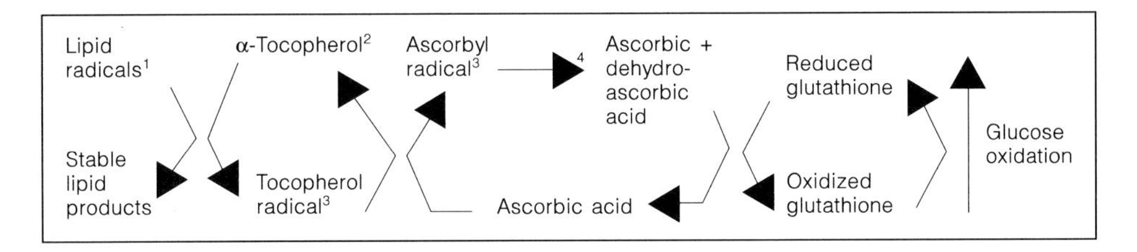

Fig. 2. Cooperation of antioxidant micronutrients and glucose in quenching reactive species and breaking chain reactions. [1] Reactive species created by free radical attack and capable of sustaining chain reactions and damaging DNA. [2] Carotenoids may act in synergism with α-tocopherol at this point. [3] Free radicals with long half-lives, slow chain reactions. [4] Reaction of two ascorbyl radicals eliminates free radical chain reaction.

Table 1. Quenching of reactive species by micronutrients

Micronutrient	Species quenched in experimental systems
Ascorbic acid	Superoxide, hydroxyl radical, hydrogen peroxide, nitrosating species (N_2O_3), hypohalous acids
Vitamin E	Hydrogen peroxide, fatty acid peroxyl and alkoxyl radicals
Carotenoids	Singlet oxygen, fatty acid peroxyl and alkoxyl radicals
Retinoids	Thiyl radical

in intake. There is, therefore, considerable interest in the ability of such manipulation to protect humans against cancer.

Experimental and Metabolic Evidence

The ability of vitamins A, C and the carotenoids to scavenge reactive species is well documented [19–25], and a listing of the species that can be quenched by these antioxidants is given in table 1. Figure 2 shows how these micronutrients can be synergistic [19, 24–32] with redox cycling and reducing power from glucose, allowing relatively small amounts of these vitamins to act out of proportion to their concentrations. The fat-soluble vitamins tend to protect cell membranes [21, 31, 33–37] rather than DNA directly. Yet, these apolar antioxidants can prevent DNA damage indirectly, because some of the reactive species (alkoxyl and peroxyl radicals and alkenals such as hydroxynonenal) generated by oxidation of polyunsaturated fatty acids in membranes can diffuse into the nucleus and cause mutagenesis [8, 19, 37, 38].

Table 2. Average concentrations of total vitamin C[1] in cells and biological fluids in health[2]

Cell type[3]	Vitamin C mmol/l	Biological fluid	Vitamin C mmol/l
Cervicovaginal	16.4[4]	Aqueous humour[6]	1.00
Monocyte/macrophage[5]	8.0	Seminal fluid	0.65
Neutrophil	1.3	Gastric juice	0.15
Brain	1.3[4]	Plasma	0.04
Gastric mucosal	2.0[4]		

[1] Ascorbic acid + dehydroascorbic acid; the latter <5% of total, except gastric juice.
[2] For sources, see Schorah [47].
[3] Values expressed per litre cell water.
[4] Estimates of concentration in cell water from wet weights.
[5] Vitamin E levels in these cells reported as 8.0 mmol/l [48].
[6] Animal studies, all other values are in man.

There are a number of studies using isolated DNA, cells in culture, or animal models which show that damage to DNA or tumourigenesis can be considerably reduced, if not eliminated, by the presence of the antioxidant micronutrients [13, 15, 39–46]. The mechanism of protection may include cell differentiation towards a normal phenotype as well as a purely antioxidant function [11, 15–17, 20]. However, are these experimental situations relevant to the prevention of carcinogenesis in man? The use of unphysiological doses of vitamins, some of which are potentially toxic [42], present a problem in these studies. In addition, animals metabolize both ascorbic acid and carotenoids in different ways, and as the effect of carcinogens can also differ from species to species, animal studies are unlikely to be of direct relevance to understanding the human carcinogenic process.

However, there is human metabolic evidence which supports the translation of these experimental findings to man. Intracellular concentrations of vitamins C and E are very high in cells where increased activity of reactive species could be expected [47] (table 2). For example, concentrations are high in mucosal cells and their secretions where environmental, chemical and infective agents or radiation could stimulate excessive reactive species generation. Such cells also undergo rapid turnover, and in these situations DNA damage is most likely to lead to permanent mutation as errors are transcribed by cell replication before they can be repaired. High antioxidant levels are also found in cells which have a high rate of oxygen consumption, such as the brain, and in cells that synthesize reactive species as part of their metabolic processes, such as phagocytic cells [47, 48].

If these antioxidant micronutrients are not involved in scavenging reactive species, what is their role in these cells? The antiscorbutic function of vitamin C in hydroxylation reactions in collagen, noradrenalin and carnitine synthesis [49] would not require such high levels in these particular cell types. Antioxidant roles have been proposed. Ascorbic acid is now known to be secreted into the gastric lumen at higher concentrations than in plasma [50, 51]. Gastric juice can provide an ideal environment for the formation of the carcinogenic N-nitroso compounds, and as ascorbate is an excellent scavenger of the nitrosating agents, its role in the gastric lumen may be to protect the stomach from these compounds [52]. Whilst the normal stomach maintains adequate concentrations for this purpose, the gastritic stomach does not [50–52], and current evidence would suggest that intakes well above those required to prevent clinical scurvy are needed in these patients to sustain reasonable concentration of ascorbic acid in their gastric juice [53, 54]. Ascorbic acid also seems to be needed at high concentrations to prevent DNA damage in human sperm [55] and workers exposed to cytostatic drugs [56].

Not only are there extremely high levels of certain antioxidants maintained in some cell types, but studies in man suggest that the metabolism of ascorbic acid may occur at greater rates than is needed to prevent clinical scurvy. The association between plasma vitamin C and intake of the vitamin is sigmoidal. Rapid increases in plasma concentration are found only when intakes rise above 40 mg/day [57, 58]. As clinical scurvy can be prevented by intakes of approximately 10 mg/day [49], this suggests utilization of the vitamin by systems independent of its antiscorbutic properties. Such additional metabolic roles could include an antioxidant function, especially as plasma values often fall to very low concentrations in critically ill patients [59] when the antioxidant defence will be most compromised by the dramatic increase in reactive species generation seen in these patients. Similar metabolic data are not yet available for the fat-soluble antioxidants, but there are reports of decreases in these micronutrients in conditions where reactive species generation is increased [60].

In general, the metabolic and experimental studies which have been undertaken would strongly support a role for antioxidant micronutrients and associated vitamins in the prevention of a variety of cancers in man, but such associations can only be confirmed or refuted by appropriate clinical studies.

Observational Studies of Cancer Prevention in Man

These studies are essentially comparisons of the antioxidant status (dietary intake or plasma concentration) in patients with and without cancer. They can be undertaken on a geographical basis, comparing regions at differential risk of malignancy, or by examining differences between cancer patients and appropriately matched controls either as the disease develops or before its onset. The best studies are the prospective studies where intakes or blood levels of these micronutrients are assessed before the disease is present, and comparisons are subsequently made between those who develop the disease and those who do not. There are over a hundred observational studies in this area, and the overall trends and findings from these studies have been summarized recently by a number of reviewers [6, 61–70]. Essentially the studies show a clear negative association between fruit and vegetable intake and cancer prevalence. The greatest protection seems to be afforded against cancers of epithelial origin (especially lung and gastrointestinal), and least against cancers which are sensitive to hormonal activity such as breast, ovary and prostate. Those individuals in the lowest quartile of fruit and vegetable intake are at 2- to 3-fold increased risk of cancer compared with subjects in the highest quartile.

When individual vitamins have been estimated, the strongest protective associations with all cancers are for carotenoids with a weak association for vitamin E. Vitamin C protection is also weak for all cancers taken together, but is strong for upper gastrointestinal, especially stomach. Carotenoids, as well as showing general protection against cancer, show a strong protective association for lung cancer. Surprisingly, the carotenoid effect seems to be independent of the potential for some carotenoids to be converted to retinol [20, 65, 68] and vitamin A appears to be only weakly associated with reduced cancer prevalence. However, because plasma levels of vitamin A are well controlled compared with those of the other antioxidant micronutrients, this may mean that neither intake nor blood levels are closely associated with protective effects of retinol within the cells. In addition, the actual conversion of β-carotene to vitamin A at the tissues may be more important than the potential for that conversion as measured by total β-carotene intake.

Prospective studies [65, 71–76] have tended to show weaker associations, but essentially the same trends are found, with the same micronutrients protecting at the same sites. One of the best prospective studies, the Basel study [72], measured blood concentration of micronutrients at the time of sampling. Again the findings reflect the trends already described for other observational studies, with β-carotene blood levels showing the closest inverse association to all cancers, especially lung and stomach, with weaker trends for vitamin C and E.

There are problems with the interpretation of all observational studies [20, 65, 76]. These include: changes in the measured parameters produced by the presence of disease affecting the cases but not the controls; sample deterioration during long periods of storage in prospective studies; recall error in retrospective studies; vitamin reserves greater than the threshold where an effect would be expected (this was believed to have occurred for vitamin E in the Basel study [73]) and associations identified which are noncausal (e.g., other factors in fruit rather than the antioxidant vitamins, are the true protective agents). However, the fact that some of these problems would spuriously strengthen an association whilst others would weaken it, and the observations that the vitamin/cancer associations tend to remain even when adjustment is made for possible confounding factors, suggest that the associations observed are real. What cannot be determined by observational studies however adjusted, is whether the observed associations are causal. This can only be confirmed by appropriately randomized and controlled intervention trials.

Intervention Studies in Man

There have been several attempts to reverse conditions which are believed to be premalignant. These include oral leukoplakia, colorectal adenomatosis and oesophageal and breast dysplasia. There has been a consistent and fairly dramatic regression of leukoplakia with retinoids and/or carotenoids [20, 77], and in one publication by α-tocopherol [78]. Unfortunately the retinoid treatments have proved to be unacceptably toxic [42]. In contrast, micronutrient treatment of premalignant conditions at other sites [76, 79–81] has failed to elicit a significant response with the possible exception of folate treatment of dysplasia, where some improvements in cytology have been noted [18]. The most obvious conclusion is that micronutrients are ineffective at preventing carcinogenesis at most sites. However, it is possible that some of these so-called premalignant changes are not part of the process of carcinogenesis. Alternatively, and arguably more likely, such changes may represent advance lesions and here only retinoids and possibly folic acid seem able to influence the process. Such late interventions fit more closely with studies which attempt to treat cancer rather than prevent the condition, evidence for which is considered in the next section.

Although there are many randomized intervention studies of cancer prevention in high-risk groups in progress at the present time, only two have

Table 3. Findings of randomized intervention studies of cancer prevention using micronutrients

Location	Supplement dose/day	Plasma β-carotene concentration, μmol/l pre-/postsupplement	Outcome
China [82]	β-carotene (15 mg) vitamin E (30 mg) selenium (50 μg)	0.11/1.6	mortality decreased: all cancers 13% stomach 21%
Finland [83]	β-carotene (20 mg) vitamin E (50 mg)	0.34/5.6	lung cancer incidence increased 18% with β-carotene: no interaction with vitamin E

reported [82, 83] (table 3). The results are conflicting. The study in China [82] has suggested that a combination of β-carotene, vitamin E and selenium is preventive, particularly for gastric cancer. In contrast, the Finnish study [83] suggests a promotional role for β-carotene especially in lung cancer, and no interaction with vitamin E. Both studies were in high-risk groups, one geographical, the other in smokers. However, the micronutrient status at baseline was much better in the Finnish group than in the Chinese population, and the levels of vitamin supplementation were also higher in the former study. The result of this was that plasma levels of β-carotene were considerably greater in those supplemented in Finland than those in China and exceeded average levels found in the plasma of well-fed populations by over 10-fold. We might, therefore, conclude that the use of β-carotene supplements in a population where reserves are already good is of little benefit, and might actually cause harm. In contrast, rectifying suboptimal levels is advantageous. This suggestion would fit with the observational studies, where the lower quartile of intake of foods rich in antioxidant vitamins is most associated with increased cancer risk. Indeed, the baseline diets in the Finnish study once more indicated that foods containing antioxidants were associated with low risk of developing lung cancer. However, the strongest epidemiological evidence is for protection by a high fruit and vegetable intake rather than specific micronutrients. The failure of a protective effect in the Finnish study may be due to the fact that the effective ingredients in fruit and vegetables were not used in this trial.

Treatment of Existing Cancers

Animal studies and especially culture of tumour cell lines have indicated that ascorbic acid, retinoids, carotenoids and tocopherols could be effective in preventing tumour cell growth and metastatic spread [11, 15, 16, 41, 42, 84–87]. Certain types of tumour cell lines have proved more susceptible to this treatment than others [88]. However, the doses used in these studies have often been very high, and potentially toxic. It is possible, therefore, that the effects seen in many of these experiments do not reflect an antioxidant role of the vitamin. Retinoids are known to be toxic to both animals and man at high doses [42], and ascorbic acid and vitamin E, especially in the presence of transition metals such as iron and copper, are capable of generating free radicals and oxidizing tissues rather than acting as scavengers of these species [30, 89–93]. In vitro systems often use media composition which would encourage the presence of traces of free iron and copper, and in some systems effects on tumour cell lines have been enhanced by the addition of these metals.

Results of treatment of existing malignancy in man have been very variable. The most convincing protective effect is in promyelocytic leukaemia [17, 94] where both carotenoids and retinoids have been found to be effective. The role of these micronutrients and possibly folic acid [18] at reversing specific chromosomal abberation needs further study.

Cameron [95] and others (see Basu and Schorah [49]) have claimed success with very high dose vitamin C regimens in advance malignancy. Randomized studies, which themselves have been problematical [96], have not confirmed these findings [49]. At best the response is that high dose vitamin C could be an adjuvant to other therapies with the response limited to checking the growth and spread of the malignancy and possibly some improvement in well-being. The very high doses used, could again suggest a toxic effect of the vitamin to which some types of cancer cells may be more susceptible.

Conclusions and Appropriate Intakes of Antioxidants and Related Micronutrients for Cancer Prevention

Currently there is little evidence that vitamins of any type are able to greatly modify the progression of established malignancy with the exception of promyelocytic leukaemia. In contrast, there is considerable laboratory evidence from chemical, cell culture and animal studies that antioxidant vitamins and related micronutrients are able to slow, or possibly prevent the carcinogenic process. There is a good theoretical basis for these findings. Current theories for the mechanism of tumourigenesis suggest that reactive species and prooxidants promote and encourage the process whilst antioxidants are inhibitory and protective. Retinoids and folate with limited or no antioxidant activity may protect DNA in other ways.

In man there is support for this role from the extraordinarily high concentrations of ascorbate and possibly α-tocopherol at sites where oxidant stress is likely to be most intense, with loss of such antioxidant protection in some conditions which predispose to malignancy. There is also impressive epidemiological agreement, particularly from observational studies, where the lowest fruit and vegetable intake has been consistently associated with increased risk of cancer, especially of the lung and gastrointestinal tract, but much less evidence that such low intakes can encourage the development of cancers which are under hormonal control. Where individual micronutrients have been considered, β-carotene appears to have the strongest protective effect followed by vitamin C and vitamin E. Whilst the experimental studies have suggested a role for retinoids, this has not been confirmed by the observational studies. Unfortunately, with the exception of oral leukoplakia, studies investigating reversal of premalignant conditions have been disappointing, and two intervention studies aimed at prevention in large populations have produced conflicting results.

All this begs the question as to what dietary advice or intervention, if any, should be provided prior to the publication of the many randomized intervention studies that are presently investigating the role of micronutrients in cancer prevention. Gey [73] has produced recommendations for minimum blood concentrations and intakes of antioxidant micronutrients. Table 4 compares these guidelines with average intakes in the United Kingdom, and the percentage of the UK population who failed to reach these recommended minimum blood levels for antioxidant micronutrients. The blood levels proposed seem appropriate, as in general they seek to maintain individuals in the upper quartile of the reference range for these micronutrients, an aim which is supported by much of the observational work. The suggested intakes to sustain these levels in the plasma also seem appropriate, with the exception of vitamin E where the intake is unnecessarily high. In table 4, I have made some alternative suggestions for intakes based not only on the observational studies in man, but also on the metabolic studies of vitamin utilization.

These recommendations can be used to encourage what current evidence suggests is a diet capable of decreasing cancer risk at some sites by about 2- to 3-fold. Individuals should only be advised to move towards the recommendations included in table 4 by dietary change and not by the use of vitamin supplements because of the findings of the Finnish study showing a promotional effect of high intakes of β-carotene supplements on lung cancer. Such

Table 4. Plasma thresholds and intakes of antioxidant micronutrients associated with decreased risk of carcinogenesis

Micronutrient	Plasma threshold[1] μmol/l	% of UK population less than threshold[2]	Suggested intake mg/d[3]		Average UK intake, mg/day[4]
			Gey[1]	Schorah	
Vitamin A	2.8	88	1.0	1.3	1.2
β-Carotene	0.45	74	6.0	5.0	2.3
Vitamin C	50.0	22[5]	60–250	60	74
Vitamin E	30.0	80	40–60	15	10.1

[1] Source, see Gey [73].
[2] Sources, Schorah [unpubl. data] and Gregory et al. [100].
[3] Estimated as intakes required to reach threshold in healthy subjects.
[4] Includes supplements [100].
[5] Healthy young: healthy elderly 56%, sick elderly 100% [101].

dietary modification should also serve to eliminate the present lowest quartile of antioxidant intake which a review of the literature suggests is associated with the greatest risk. These suggested changes in intake could be achieved by a more Mediterranean-style diet which for many will include increased consumption of fruit and vegetables (found to be significantly protective in observational studies), olive oil and fish (provides a high vitamin E/polyunsaturated fat ratio, believed to protect from ischaemic heart disease [97, 98]) at the expense of animal proteins and saturated fats. Such a diet could, of course, include the occasional glass of red wine. The suggestion of 5 fruits-vegetables/day/person is a similar recommendation, but with less than 10% of the US population reaching this intake [99] this may be a less palatable option.

References

1 Weinberg RA: Oncogens, antioncogenes and the molecular basis of multistep carcinogenesis. Cancer Res 1989;49:3713–3721.
2 Sun Y: Free radicals, antioxidant enzymes and carcinogenesis. Free Radic Biol Med 1990;8:583–599.
3 Guyton KZ, Kensler TW: Oxidative mechanism in carcinogenesis. Br Med Bull 1993;49:523–544.
4 Dizdaroglu M: Chemistry of free radical damage to DNA; in Halliwell B, Aruoma OI (eds): DNA and Free Radicals. Chichester, Ellis Horwood, 1993, pp 19–39.
5 Ramotar D, Demple B: Enzymes that repair oxidative damage to DNA; in Halliwell B, Aruoma OI (eds): DNA and Free Radicals. Chichester, Ellis Horwood, 1993, pp 163–191.
6 Ames BN, Shigenaga MK: Oxidants are a major contributor to cancer and ageing; in Halliwell B, Aruoma IO (eds): DNA and Free Radicals. Chichester, Ellis Horwood, 1993, pp 1–15.

7 Winrow VR, Winyard PG, Morris CJ, Blake DR: Free radicals in inflammation: Second messengers and mediators of tissue destruction. Br Med Bull 1993;49:506–522.
8 Meheghini R, Martins EL: Hydrogen peroxide and DNA damage; in Halliwell B, Aruoma OI (eds): DNA and Free Radicals. Chichester, Ellis Horwood, 1993, pp 83–93.
9 Halliwell B: Oxidative DNA damage: Meaning and measurement; in Halliwell B, Aruoma OI (eds): DNA and Free Radicals. Chichester, Ellis Horwood, 1993, pp 67–69.
10 Bergelson S, Pinkus R, Daniel V: Intracellular glutathione levels regulate Fos/Jun induction and activation of glutathione s-transferase gene expression. Cancer Res 1994;54:36–40.
11 Prasad KN, Edwards-Prasad J, Kumar S, Meyers A: Vitamins regulate gene expression and induce differentiation and growth inhibition of cancer cells. Arch Otolaryngol Head Neck Surg 1993;119: 1133–1140.
12 Lunec J: Free radicals: Their involvement in disease processes. Ann Clin Biochem 1990;27:173–182.
13 Chojkier M, Honglum K, Solis-Herruzo J, Brenner DA: Stimulation of collagen gene expression by ascorbic acid in cultured human fibroblasts. A role for lipid peroxidation. J Biol Chem 1989; 264:16957–16962.
14 Geesin JC, Hendricks LJ, Falkenstein PA, Gordon JS, Berg RA: Regulation of collagen synthesis by ascorbic acid: Characterisation of the role of ascorbate-stimulated lipid peroxidation. Arch Biochem Biophys 1991;290:127–132.
15 Bollag W, Hartmann HR: Prevention and therapy of cancer with retinoids in animals and in man. Cancer Surv 1983;2:293–314.
16 Poddar S, Hong WK, Thacher SM, Lotan R: Retinoic acid suppression of squamous differentiation of human head-and-neck squamous carcinoma cells. Int J Cancer 1991;48:239–247.
17 Breitman TR, Selonick SE, Collins SJ: Induction of differentiation of the human promyelocytic leukaemia cell line (HL-60) by retinoic acid. Proc Natl Acad Sci USA 1980;77:2936-2940.
18 Butterworth CE: Folate deficiency and cancer; in Bendich A, Butterworth CD (eds): Micronutrients in Health and in Disease Prevention. New York, Dekker, 1991, pp 165–183.
19 Morrero R, Marnett LJ: The role of organic peroxyl radicals in carcinogenesis; in Halliwell B, Aruoma OI (eds): DNA and Free Radicals. Chichester, Ellis Horwood, 1993, pp 145–161.
20 Van Poppel G: Carotenoids and cancer: An update with emphasis on human intervention studies. Eur J Cancer 1993;29A:1335–1344.
21 Daquino M, Dunster C, Willson RL: Vitamin A and glutathione-mediated free radical damage. Biochem Biophys Res Commun 1989;161:1199–1203.
22 Machlin LJ, Bendich A: Free radical tissue damage: Protective role of antioxidant nutrients. FASEB J 1987;1:441–445.
23 Nandi A, Chatterjee IB: Scavenging of superoxide radical by ascorbic acid. J Biosci (Z Naturforsch [C]) 1987;11:435–441.
24 Niki E: Action of ascorbic acid as a scavenger of active and stable oxygen radicals. Am J Clin Nutr 1991;54:1119S–1124S.
25 McCay PB: Vitamin E interaction with free radicals and ascorbate. Annu Rev Nutr 1985;5:323–340.
26 Liebler DC, Kling DS, Reed DJ: Antioxidant protection of phospholipid bilayer by α-tocopherol: Control of α-tocopherol status and lipid peroxidation by ascorbic acid and glutathione. J Biol Chem 1986;261:12114–12119.
27 Liebler DC, Kaysen KL, Kennedy TA: Redox cycles of vitamin E: Hydrolysis and ascorbic acid-dependent reduction of 8α-(alkyldioxy)tocopherones. Biochemistry 1989;28:9772–9777.
28 Sato K, Niki E, Shimasaki H: Free radical-mediated chain oxidation of low-density lipoprotein and its synergistic inhibition by vitamin E and vitamin C. Arch Biochem Biophys 1990;279:402–405.
29 Scarpa N, Rigo A, Maiorino M, Orsini F, Gregolin C: Formation of α-tocopherol radical and recycling of α-tocopherol by ascorbate during peroxidation of phosphatidylcholine liposomes. Biochim Biophys Acta 1984;801:215–219.
30 Stadtman ER: Ascorbic acid and oxidative inactivation of proteins. Am J Clin Nutr 1991;54: 1125S–1128S.
31 Buettner GR: The pecking order of free radicals and antioxidants: Lipid peroxidation, α-tocopherol and ascorbate. Arch Biochem Biophys 1993;300:535–546.

32 Palozza P, Krinsky NI: β-Carotene and α-tocopherol are synergistic antioxidants. Arch Biochem Biophys 1992;297:184–187.
33 Urano S, Hoshi-Hashizume M, Tochigi N, Matsuo M, Shiraki M, Ito H: Vitamin E and the susceptibility of erythrocytes and reconstituted liposomes to oxidative stress in aged diabetics. Lipids 1991;26:58–61.
34 Frei B, Forte TM, Ames BN, Cross CE: Gas phase oxidants of cigarette smoke induce lipid peroxidation and changes in lipoprotein properties in human blood plasma-protective effects of ascorbic acid. Biochem J 1991;277:133–138.
35 Harats D, Bennaim M, Dabach Y, Hollander G, Hauivi E, Stein O, Stein Y: Effect of vitamin C and vitamin E supplementation on susceptibility of plasma lipoproteins to peroxidation induced by acute smoking. Atherosclerosis 1990;85:47–54.
36 Jialal I, Vega GL, Grundy SM: Physiological levels of ascorbate inhibit the oxidative modification of low density lipoproteins. Atherosclerosis 1990;82:185–191.
37 Cheeseman KH: Lipid peroxidation and cancer; in Halliwell B, Aruoma OI (eds): DNA and Free Radicals. Chichester, Ellis Horwood, 1993, pp 109–144.
38 Siems W, Zollner H, Grune T, Esterbauer H: The metabolism of 4-hydroxynonenal in liver and small intestine, an antioxidative defense system. Int J Radiat Biol 1992;62:116.
39 Yasukawa M, Terasima T, Seki M: Radiation-induced neoplastic transformation of C3HIOT$\frac{1}{2}$ cells is suppressed by ascorbic acid. Radiat Res 1989;120:456–467.
40 Lupulescu A: Control of epithelial precancer cell transformation into cancer cells by vitamins. J Cell Biol 1986;103:29A.
41 Lupulescu A: Inhibition of DNA synthesis and neoplastic cell growth by vitamin A (retinol). J Natl Cancer Inst 1986;77:149–156.
42 Greenberg ER: Retinoids or carotenoids: Is there another choice. Prev Med 1993;22:723–727.
43 Odeleye OE, Eskelson CD, Mufti SI, Watson RR: Vitamin E inhibition of lipid peroxidation and ethanol mediated promotion of esophageal tumorigenesis. Nutr Cancer 1992;17:223–234.
44 Shklar G: Oral mucosal carcinogenesis in hamsters: Inhibition by vitamin E. J Natl Cancer Inst 1982;68:791–797.
45 Santamaria L, Bianchi A: Cancer chemoprevention by supplemental carotenoids in animals and humans. Prev Med 1989;18:603–623.
46 Van Staden AM, Van Rensburg CE, Anderson R: Vitamin E protects mononuclear leucocyte DNA against damage mediated by phagocyte-derived oxidants. Mutat Res 1993;288:257–262.
47 Schorah CJ: The transport of vitamin C and effects of disease. Proc Nutr Soc 1992;51:189–198.
48 Pacht ER, Kaseki H, Mohammed JR, Cornwell DG, Davis WB: Deficiency of vitamin E in the alveolar fluid of cigarette smokers. J Clin Invest 1986;77:789–796.
49 Basu TK, Schorah CJ: Vitamin C in Health and Disease. London, Croom Helm, 1982.
50 Sobala GM, Schorah CJ, Sanderson M, Dixon MF, Tomkins DS, Godwin P, Axon ATR: Ascorbic acid in the human stomach. Gastroenterology 1989;97:357–363.
51 Sobala GM, Pignatelli B, Schorah CJ, Bartsch H, Sanderson M, Dixon MF, Shires SM, King RFG, Axon ATR: Levels of nitrite, N-nitroso compounds, ascorbic acid and total bile acids in gastric juice of patients with and without precancerous conditions of the stomach. Carcinogenesis 1991;12:193–198.
52 Schorah CJ, Sobala GM, Sanderson M, Collis N, Primrose JN: Gastric juice ascorbic acid: Effects of disease and implications for gastric carcinogenesis. Am J Clin Nutr 1991;53:287S–293S.
53 Sobala GM, Schorah CJ, Shires S, Lynch DAF, Gallacher B, Dixon MF, Axon ATR: Effect of eradication of *Helicobacter pylori* on gastric juice ascorbic acid concentrations. Gut 1993;34:1038–1041.
54 Waring AJ, Drake I, Schorah CJ, White KLM, Lynch DAF, Axon ATR, Dixon MF: Ascorbic acid and total vitamin C concentrations in plasma, gastric juice and gastrointestinal mucosa: Effects of gastritis and oral supplementation. Gut 1994;35(suppl):S19.
55 Fraga CG, Motchnik PA, Shigenaga MK, Helbock HJ, Jacob RA, Ames BN: Ascorbic acid protects against endogenous oxidative DNA damage in human sperm. Proc Natl Acad Sci USA 1992;88:11003–11006.

56 Rössner P, Cerna M, Pokorna D, Hajek V, Petr J: Effect of ascorbic acid prophylaxis on the frequency of chromosome aberrations, urine mutagenicity and nucleolus test in workers exposed to cytostatic drugs. Mutat Res 1988;208:149–153.
57 Newton HMV, Morgan DB, Schorah CJ, Hullin RP: Relation between intake and plasma concentration of vitamin C in elderly women. Br Med J 1983;287:1429.
58 Newton HMV, Schorah CJ, Habibzadeh N, Morgan DB, Hullin RP: The cause and correction of low blood vitamin C concentrations in the elderly. Am J Clin Nutr 1985;42:656–659.
59 Downing C, Piripitsi A, Bodenham A, Schorah CJ: Plasma vitamin C concentrations in critically ill patients. Proc Nutr Soc 1993;52:314A.
60 Goode HF, Cowley H, Leek JP, Webster NR: Antioxidant vitamin status, lipid peroxidation, and indices of nitric oxide production in patients with sepsis and secondary organ dysfunction. Proc Nutr Soc 1993;52:334A.
61 Chen LH, Boissonneault GA, Glavert HP: Vitamin C, vitamin E and cancer. Anticancer Res 1988; 8:739.
62 Block G: Vitamin C and cancer prevention: The epidemiological evidence. Am J Clin Nutr 1991; 53:270S–282S.
63 Block G, Patterson B, Subar A: Fruit, vegetables and cancer prevention: A review of the epidemiological evidence. Nutr Cancer 1992;18:1–29.
64 Zeigler RG: A review of epidemiologic evidence that carotenoids reduce the risk of cancer. J Nutr 1989;119:116–122.
65 Zeigler RG: Vegetables, fruit and carotenoids and risk of cancer. Am J Clin Nutr 1991;53:251S–259S.
66 Dorgan JF, Schatzkin A: Antioxidant micronutrients and cancer prevention. Hematol Oncol Clin North Am 1991;5:43–68.
67 Gridley G, McLaughlin JK, Block G, Blot WJ, Gluch M, Fraumeni JF: Vitamin supplement use and reduced risk of oral and pharyngeal cancer. Am J Epidemiol 1992;135:1083–1092.
68 Mayne ST, Janerick DT, Greenwald P, Chorost S, Tucci C, Zamen MB, Melamed MR, Kiely M, McKneally MF: Dietary β-carotene and lung cancer risk in US nonsmokers. J Natl Cancer Inst 1994;86:33–38.
69 Barone J, Taiole E, Herbert JR, Wynder EL: Vitamin supplement use and risk for oral and oesophageal cancer. Nutr Cancer 1992;18:31–41.
70 Fontham ETH, Pickle LW, Haenszel W, Correa P, Lin Y, Falk RT: Dietary vitamins A and C and lung cancer risk in Louisiana. Cancer 1988;62:2267–2273.
71 Connett JE, Kuller LH, Kjelsberg MO, Polk BF, Collins G, Rider A, Hulley SB: Relationship between carotenoids and cancer: the multiple risk factor intervention trial (MRFIT) study. Cancer 1989;64:126–134.
72 Stahelin HB, Gey KF, Eichholzer M, Ludin E, Bernasconi F, Thurneysen J, Brubacher G: Plasma antioxidant vitamins and subsequent cancer mortality in the 12-year follow-up of the prospective Basel study. Am J Epidemiol 1991;133:766–775.
73 Gey KF: Prospects for the prevention of free radical disease, regarding cancer and cardiovascular disease. Br Med Bull 1993;49:679–699.
74 Comstock GW, Bush TL, Helzlsover K: Serum retinol, β-carotene, vitamin E and selenium as related to subsequent cancer at specific sites. Am J Epidemiol 1992;135:115–121.
75 Zheng W, Blot WJ, Diamond EL, Norkus EP, Spate V, Morris JS, Comstock GW: Serum micronutrients and the subsequent risk of oral and pharyngeal cancer. Cancer Res 1993;53:795–798.
76 Kneckt P, Aromaa A, Maatela J, Ritua-Kaarina A, Nikkari T, Hakama H, Hakuinen T, Peto R, Teppo L: Vitamin E and cancer prevention. Am J Clin Nutr 1991;53:283S–286S.
77 Garewaal HS, Meyskens FL, Killen D, Reeves D, Kiersch TA, Elletson H, Stasberg A, King D, Steinbrann K: Response of oval leukoplakia to β-carotene. J Clin Oncol 1990;8:1715-1720.
78 Benner SE, Winn RJ, Lippman SM, Polano J, Hansen KS, Luna MA, Hong WK: Regression of oral leukoplakia with α-tocopherol: A community clinical oncology program chemoprevention study. J Natl Cancer Inst 1993;85:44–47.
79 McKeown-Eyssen G, Holoway C, Jazmaji V, Bright-See E, Dion P, Bruce WR: A randomised trial of vitamin C and vitamin E in the prevention of recurrence of colorectal polyps. Cancer Res 1988; 48:4701–4705.

80 De Cosse JJ, Miller HH, Lesser ML: Effect of wheat fibre and vitamin C and E on rectal polyps in patients with familial adenomatous polyposis. J Natl Cancer Inst 1989;81:1290–1297.
81 Paganelli GM, Biasco G, Brandi G, Santucci R, Gizzi G, Villani V, Cianci M, Miglioli M, Barbara L: Effect of vitamin A, C and E supplementation on rectal cell proliferation in patients with colorectal adenomas. J Natl Cancer Inst 1992;84:47-52.
82 Blot WJ, Li JY, Taylor PR, Guo W, Dawsey S, Wang GQ, Yang CS, Zheng SUF, Gail M, Li GY, Zu Y, Liu BQ, Tangrea J, Sun YH, Liu F, Fraumeni J, Zhang YH, Li B: Nutrition intervention trials in Linsian, China. J Natl Cancer Inst 1993;85:1483–1492.
83 The α-Tocopherol, β-Carotene Cancer Prevention Study Group: The effect of vitamin E and β-carotene on the incidence of lung cancer and other cancers in male smokers. N Engl J Med 1994; 330:1029–1035.
84 Bran S, Froussard P, Guichard M, Jasmin C, Augery Y, Sinoussi-Barre F, Wray W: Vitamin C preferential toxicity for malignant melanoma cells. Nature 1980;284:629–631.
85 Gardener NS, Duncan JR: Enhanced prostaglandin synthesis as a mechanism for inhibition of melanoma cell growth by ascorbic acid. Prostaglandins Leukotrienes Essential Fatty Acids 1988; 34:119-126.
86 Taper HS, De Gerlache J, Lans M, Roberfroid M: Non-toxic potentiation of cancer chemotherapy by combined C and K_3 vitamin pre-treatment. Int J Cancer 1987;40:575–579.
87 Lupulescu A: Vitamin C inhibits DNA, RNA and protein synthesis in neoplastic epithelial cells. Int J Vitam Nutr Res 1991;61:125–129.
88 Kao TL, Mayer WJ, Post JFM: Inhibiting effects of ascorbic acid on growth of leukemic and lymphoma cell lines. Cancer Lett 1993;70:101–106.
89 Ahmed FF, Cowan DL, Sun AY: Potentiating of ethanol-induced lipid peroxidation of biological membranes by vitamin C. Life Sci 1988;43:1169–1176.
90 Miller DM, Aust SD: Studies of ascorbate-dependent iron-catalysed lipid peroxidation. Arch Biochem Biophys 1989;271:113–119.
91 Kobayashi S, Veda K, Morita J, Sakai H, Komano T: DNA damage induced by ascorbate in the presence of Cu^{2+}. Biochim Biophys Acta 1988;949:143–147.
92 Backowski GJ, Thomas JD, Girotti AW: Ascorbate enhanced lipid peroxidation in photo-oxidised cell membranes. Lipids 1988;23:580–586.
93 Burkitt MJ, Gilbert BC: Model studies on the iron-catalysed Haber-Weiss cycle and the ascorbate-driven Fenton reaction. Free Radic Res Commun 1990;10:265–280.
94 Van der Weyden MB, Cortis DJ, Szer J: Vitamin therapy for acute leukaemia. Aust N Z J Med 1992;22:446–447.
95 Cameron E: Protocol for the use of vitamin C in the treatment of cancer. Med Hypotheses 1991; 36:190–193.
96 Richards E: The politics of therapeutic evaluation: The vitamin C and cancer controversy. Soc Stud Sci 1988;18:653–701.
97 Gey KF: Cardiovascular disease and vitamins; in Walter P (ed): The Scientific Basis for Vitamin Intake in Human Nutrition. Bibl Nutr Dieta. Basel, Karger, 1995, No 52, pp 75–91.
98 Sanders TAB: The Mediterranean diet: Fish and olives, oil on troubled waters. Proc Nutr Soc 1991; 50:513–517.
99 Patterson BH, Block G: Fruit and vegetable consumption: National survey data; in Bendich A, Butterworth CD (eds): Micronutrients in Health and in Disease Prevention. New York, Dekker, 1991, pp 409–436.
100 Gregory G, Foster K, Tyler H, Wiseman M: The Dietary and Nutritional Survey of British Adults. London, HMSO, 1992.
101 Morgan DB, Newton HMV, Schorah CJ, Jewitt MA, Hancock MR, Hullin RP: Abnormal indices of nutrition in the elderly: A study of different clinical groups. Age Ageing 1986;15:65–76.

Dr. C.J. Schorah, Division of Clinical Sciences, School of Medicine,
The Old Medical School, University of Leeds, LS2 9JT (UK)

Walter P (ed): The Scientific Basis for Vitamin Intake in Human Nutrition.
Bibl Nutr Dieta. Basel, Karger, 1995, No 52, pp 108–115

The Requirement for Vitamins in Aging and Age-Associated Degenerative Conditions

Jeffrey Blumberg

USDA Human Nutrition Research Center on Aging, Tufts University, Boston, Mass., USA

Aging is accompanied by a variety of economic, psychologic, and social changes that can compromise nutritional status. In addition, aging produces physiologic changes that affect the need for several essential nutrients. Although older persons represent a very diverse group, the 1989 US National Research Council's *Recommended Dietary Allowances* (RDA) continues to provide guidelines for assessing the intake of energy and specified nutrients for adults up to age 50 and for 51 years and above [1]. However, the physiologic functions and health status of persons who are 50–60 years old are very different from those of persons who are 80–90 years old. A growing body of scientific literature and human studies with elderly subjects provides ample evidence today for altered nutrient requirements with aging [2].

Defining the changing nutrient requirements with age is particularly important as diet now presents itself as a key part of the solution to the demographic challenge of the growing population of older adults and the public health policy necessary to reduce chronic diseases. Nutritional status surveys of the elderly have shown a low-to-moderate prevalence of frank nutrient deficiencies but a marked increase in risk of malnutrition and evidence of subclinical nutrient deficiencies. The significance of these observations becomes clear with the recognition that nutritional status influences the age-related rate of functional decline in many organ systems. Nutrition is also an important factor in the progressive changes in body composition associated with aging such as the loss of bone and lean body mass. Moreover, the evidence is now undisputed that diet and nutrition are directly linked to many of the chronic diseases afflicting older adults and the elderly. Consideration of dietary recommendations for older adults must recognize each of these aspects of the relationship between nutrition and aging [3].

Vitamin Requirements during Aging

Vitamin D

Aging decreases by greater than 2-fold the capacity of the skin to produce provitamin D_3 (7-dehydrocholesterol). Older adults are also less able to synthesize 1,25-dihydroxyvitamin D in the kidney under the influence of parathyroid hormone although increasing 25-hydroxyvitamin D intake, the substrate for the renal 1α-hydroxylase, does enhance production of the vitamin. Small seasonal increases in serum parathyroid hormone levels, indicative of vitamin D depletion and with potential adverse consequences on bone-mineral balance, are apparent in healthy elderly [4]. Daily vitamin D intakes of 5.5 to over 12 μg are required in individual postmenopausal women to maintain constant serum 25-hydroxyvitamin D and parathyroid hormone concentrations throughout the year [5].

Vitamin B_6

A high prevalence of marginal vitamin B_6 deficiency in the elderly is reflected by their reduced activity of aspartate and alanine aminotransferases, increased excretion of xanthurenic acid following a tryptophan load, and lowered levels of serum pyridoxal and plasma total vitamin B_6 [6]. Ribaya-Mercado et al. [7] and Meydani et al. [8] reported that intakes >2 mg/day are necessary to normalize several parameters of vitamin B_6 status, including enzyme activity, tryptophan metabolism, immune responsiveness, and plasma vitamers, in healthy older adults.

Riboflavin

There is a marked prevalence of low vitamin B_2 intake and poor riboflavin status among elderly in many populations [9]. A metabolic study by Boisvert et al. [10] has demonstrated that riboflavin requirements are unchanged by age and suggests they be increased from current standards. Further suggesting an increased requirement for riboflavin is a study by Winters et al. [11] showing that exercise training in older women increases erythrocyte glutathione reductase activity and decreases urinary riboflavin excretion.

Vitamin B_{12}

There is an age-related trend for declining serum vitamin B_{12} within the normal range possibly due to undetected cases of pernicious anemia and/or atrophic gastritis with cobalamin malabsorption [12, 13]. Elevated serum concentrations of methylmalonic acid and homocysteine in the presence of 'low-but-normal' levels of vitamin B_{12} suggest a marked prevalence of subclinical vitamin B_{12} deficiency among the older people which may be associated with several neuropsychiatric disorders [14, 15]. Thus, Russell and Suter [2] have suggested it was imprudent to have lowered the 1989 RDA for vitamin B_{12}.

Folic Acid

Although folate intakes at the RDA level for older adults are just adequate to maintain folate nutriture (as assessed by whole blood folate), this recommended allowance does not account for folate malabsorption in the substantial number of older people with atrophic gastritis or the risk of cervical dysplasia associated with low folate status in tissue (even when blood folate values are normal) [13, 16]. Poor folate (and vitamin B_{12}) status can also result in elevated homocysteine levels which appear to be an independent risk factor in atheromatous disease [17].

Other Vitamins

Age-related decreases in vitamin K (expressed as phylloquinone/triglyceride) suggest that older adults may have altered dietary requirements to maintain an adequate status of this fat-soluble vitamin [18]. Vitamins C and E and β-carotene inhibit oxidative injury to lipids, proteins and nucleic acids; these and other mechanisms unrelated to their antioxidant actions may be responsible for the association of intakes higher than current recommended allowances for these nutrients with a reduced risk of cancer, heart disease, and infectious diseases [19].

Vitamins and Age-Associated Degenerative Conditions

Chronic Respiratory Conditions

Proteolytic and oxidant injury may be important in the pathogenesis of accelerated loss of lung function with age and disease. Higher dietary intake and serum concentrations of vitamin C have been shown to protect against the development of chronic respiratory symptoms like bronchitis and wheezing [20]. Schwartz and Weiss [21] analyzed the results of 2,526 spirometric examinations conducted as part of the NHANES I and found positive correlation between dietary vitamin C intake and pulmonary function (forced expiratory volume in 1 s). This relationship was dose–dependent with a 40-ml difference noted between the lowest (17 mg/day) and highest (178 mg/day) tertile of vitamin C intake, a magnitude roughly equivalent to 1 year of aging. If this effect is cumulated over 20–30 years, it could have a physiologically meaningful impact on the rate of decline of pulmonary function. The positive influence of vitamin C on pulmonary function appears applicable to all subjects examined including smokers and those with asthma and bronchitis. While acute administration of vitamin C causes no short-term change in pulmonary function or responsiveness of airways, some studies suggest that regular consumption of 500–1,000 mg/day may have a protective effect in asthmatic subjects [22, 23]. The mechanism underlying this relationship may be related to the protection of α_1-protease inhibitor by vitamin C against the oxyradicals generated by activated neutrophils in lung airways and parenchyma, modulation of leukotriene production, and/or enhancement of immune function. Interestingly, peak expiratory flow rate has previously been reported as a predictor of total mortality and cardiovascular mortality and as a correlate of cognitive and physical function in older populations [24].

Degenerative Eye Diseases

Oxidation of lens proteins appears to play a central role in the formation of senile cataract; over decades, UV photo-oxidative denaturation of crystallins eventually overcomes age-related declines in antioxidant and proteolytic defenses in the lens resulting in protein precipitation and opacification [25]. Mostly all the epidemiological studies and two clinical trials completed to date suggest that increased intake of micronutrients, expecially those with antioxidant capacity, are associated with a significant reduction in risk of cataract [26]. However, inconsistencies exist between these studies with regard to specific vitamins, type of cataract, and magnitude of protection; study characteristics including geographic location, method of nutritional assessment, clinical endpoint, and subject selection may explain some of these discrepancies. Supporting these observations, moderate use of daily supplemental multivitamins or riboflavin (3 mg) and niacin (40 mg) [27] and vitamins C ($\geq$300 mg) and E ($\geq$400 mg) [28] have been associated with reduced risk of cataract. The efficacy of riboflavin in the Linxian, China trials [27] may be based on its conversion to FAD, a cofactor for glutathione reductase.

Age-related macular degeneration (AMD) may be associated with the retina's high susceptibility to oxidative stress due to the high polyunsaturated fatty acid content of photoreceptor outer-segment membranes and their proximity to visible radiation. A large case-control study [29] and a prospective, longitudinal study [30] indicate that increased intakes of vitamin E or carotenoids with contributing actions of vitamin C and selenium are associated with a significantly lower risk of AMD. Limited evidence also suggests that increasing antioxidant nutrient status via supplementation may slow vision loss in patients with AMD [31].

Diabetes

Glucose intolerance and noninsulin-dependent (type II) diabetes are common among older, especially overweight, adults. Antioxidant vitamin status may be a relevant risk factor as poor metabolic control in diabetes appears to be partly a consequence of lipid peroxidation events [32] and oxidative stress appears to be involved in the genesis of diabetic complications [33]. In a randomized, crossover, and double-blind study, Paolisso et al. [34] found that daily supplementation with *dl*-α-tocopheryl acetate (900 mg) for 4 months improved insulin action not only in type II diabetic patients but also in healthy, middle-aged control subjects. They reported the area under the curve for glucose was significantly lower after vitamin E administration than during the placebo period and that both total-body glucose disposal and nonoxidative glucose metabolism were improved with the increased intake of vitamin E. Paolisso et al. [35] also found that vitamin E supplementation improves metabolic control (but not insulin secretion) in elderly type II diabetic patients. These and other studies have also demonstrated that vitamin E enhances glutathione concentrations in plasma and red blood cells; changes in the ratio of oxidized:reduced glutathione in plasma may affect β-cell responses to glucose [36].

Hypertension

The development of hypertension is influenced by genetic, hormonal, and nutritional factors. Trout [37] reviewed seven recent epidemiological studies from the United States, Finland, and Japan involving over 12,000 subjects noting that ascorbic acid intake and/or plasma concentrations were inversely correlated with both systolic and diastolic blood pressure. Moran et al. [38] found that plasma ascorbic acid concentrations were inversely correlated with blood pressure in normotensive and asymptomatic hypertensive subjects following their usual diets; the mean blood pressure in subjects in the lowest quintile of plasma ascorbic acid was significantly (5 mm Hg) higher than those in the highest quintile. Small intervention studies using vitamin C have yielded various effects on blood pressure, but Koh [39] reported daily doses of vitamin C (1,000 mg) reduced blood pressure in mildly hypertensive but not normotensive women. Vitamin C has been postulated to influence blood pressure via modulation of prostacyclin and leukotriene metabolism, lowering the sodium content in blood, and/or recycling tocopherol radicals [37].

Parkinson's Disease (PD)

A variety of oxidative mechanisms involving the activity of monamine oxidase and the formation of free radicals have been implicated in the normal age-related loss of dopaminergic neurons in the substantia nigra and in the pathogenesis of PD [40]. Thus, it should be theoretically possible to slow the progression of the disease by reducing the formation or increasing the scavenging of oxyradicals in the brain. Fahn [41] reported in uncontrolled pilot studies that daily supplementation with vitamin E (3,200 IU) and vitamin C (3,000 mg) in 15 patients with early or mild PD (and receiving anticholinergics and amantadine) delayed the need to initiate levodopa therapy by 32–40 months relative to a group not receiving antioxidants. The multicenter controlled clinical trial, Deprenyl and Tocopherol Antioxidative Therapy of Parkinsonism (DATATOP), employed daily doses of deprenyl (10 mg), a selective and irreversible inhibitor of type B monoamine oxidase, and/or racemic *dl*-α-tocopherol (2,000 IU), and/or placebo in 800 patients with stage I or II PD [42]. The deprenyl treatment delayed by 9 months the onset of PD disability prompting the clinical decision to begin administering levodopa; however, vitamin E was without effect on this primary endpoint of the trial when given alone and did not improve the results when combined with deprenyl. The investigators note that the failure of vitamin E to influence the progression of PD in this study does not preclude the potential effectiveness of different doses and/or other antioxidants.

It is noteworthy that high doses of vitamin E have been found effective in the treatment of tardive dyskinesia, a neurologic disorder characterized by choreoathetoid and other types of abnormal involuntary movement which occurs late in the course of neuroleptic treatment [43]. While the pathophysiology of tardive dyskinesia is largely unknown, altered dopamine function and degeneration of brain stem neuronal pathways by free radicals have been proposed as potential mechanisms [44].

Conclusion

Consideration of the new evidence for changed nutrient requirements associated with aging raises the important issue of defining appropriate criteria for the selection of recommended vitamin intakes. Many of the criteria currently employed to establish dietary standards lack the sensitivity to detect subtle nutrition-sensitive alterations in metabolism with significant consequences for the aging process or place little weight on the risk factors of chronic diseases common among the elderly. It now appears possible to determine optimal levels of physiologic function for older age groups and design the nutrient intakes to achieve them. Further, the allowances for vitamins can now focus on intakes which not only prevent deficiency states but are associated with maximal risk reduction of chronic disease and disability.

References

1 Food and Nutrition Board: Recommended Dietary Allowances, ed 10. Commission on Life Sciences, National Research Council. Washington, National Academy Press, 1989.
2 Russell RM, Suter PM: Vitamin requirements of elderly people: An update. Am J Clin Nutr 1993; 58:4–14.
3 Blumberg JB: Considerations of the recommended dietary allowances for older adults. Clin Appl Nutr 1991;1:9–18.
4 Lips P, Hackeng WH, Jongen MJ, van Ginkel FC, Netelenbos JC: Seasonal variation in serum concentrations of parathyroid hormone in elderly people. J Clin Endocrinol Metab 1983;57:204–206.
5 Krall EA, Sahyoun N, Tannenbaum S, Dallal GE, Dawson-Hughes B: Effect of vitamin D intake on seasonal variations in parathytoid hormone secretion in postmenopausal women. N Engl J Med 1989;321:1777–1783.
6 Kant AK, Moser-Veillon PB, Reynolds RD: Effect of age on changes in plasma, erythrocyte, and urinary B_6 vitamers after an oral vitamin B_6 load. Am J Clin Nutr 1988;48:1284–1290.
7 Ribaya-Mercado JD, Russell RM, Sahyoun N, Morrow FD, Gershoff SN: Vitamin B_6 requirements of elderly men and women. J Nutr 1991;121:1062–1074.
8 Meydani SN, Ribaya-Mercado JD, Russell RM, Sahyoun N, Morrow FD, Gershoff SN: Vitamin B_6 deficiency impairs interleukin-2 production and lymphocyte proliferation in elderly adults. Am J Clin Nutr 1991;53:1275–1280.
9 Porrini M, Testolin G, Simonetti P, Moneta A, Rovati P, Aguzzi F: Nutritional status of non-institutionalized elderly people in North Italy. Int J Vitam Nutr Res 1987;57:203–216.
10 Boisvert WA, Mendoza I, Castañeda C, De Portocarrero L, Solomons NW, Gershoff SN, Russell RM: Riboflavin requirement of healthy elderly humans and its relationship to macronutrient composition of the diet. J Nutr 1993;123:915–925.
11 Winters LRT, Yoon JS, Kalwarf HJ, Davies JC, Berkowitz MG, Haas J, Roe DA: Riboflavin requirements and exercise adaptation in older women. Am J Clin Nutr 1992;56:526–532.
12 Nilsson-Ehle H, Jagenburg R, Landahl S, Lindstedt S, Svanborg A, Westin J: Serum cobalamins in the elderly: A longitudinal study of a representative population sample from age 70 to 81. Eur J Haematol 1991;47:10–16.
13 Kassarjian Z, Russell RM: Hypochlorhydria: A factor in nutrition. Annu Rev Nutr 1989;9:271–284.
14 Selhub J, Jacques PF, Wilson PWF, Rush D, Rosenberg IH: Vitamin status and intake as primary determinants of homocysteinemia in an elderly population. JAMA 1993;270:2693–2698.
15 Beck WS: Neuropsychiatric consequences of cobalamin deficiency. Adv Intern Med 1991;36:33–56.
16 Butterworth CE Jr, Hatch KD, Gore H, Mueller H, Krumdieck CL: Improvement in cervical dysplasia associated with folic acid therapy in users of oral contraceptives. Am J Clin Nutr 1982; 35:73–82.
17 McCully KS, Olszewski AJ, Vezeridis MP: Homocysteine and lipid metabolism in atherogenesis: Effect of the homocysteine thiolactonyl derivatives, thioretinaco and thioretinamide. Atherosclerosis 1990;83:197–206.
18 Sadowski JA, Hood SJ, Dallal GE, Garry P: Phylloquinone in plasma from elderly and young adults: Factors influencing its concentration. Am J Clin Nutr 1989;50:100–108.
19 Slater TF, Block G: Antioxidant vitamins and β-carotene in disease prevention. Am J Clin Nutr 1991;53:189S–396S.
20 Schwartz J, Weiss ST: Dietary factors and their relationship to respiratory symptoms: NHANES II. Am J Epidemiol 1990;132:67–76.
21 Schwartz J, Weiss ST: Relationship between dietary vitamin C intake and pulmonary function in the First National Health and Nutrition Examination Survey (NHANES I). Am J Clin Nutr 1994; 59:110–114.
22 Schachter EN, Schlesinger A: The attenuation of exercise induced bronchospasm by ascorbic acid. Ann Allergy 1982;49:146–151.
23 Mohsenin V, DuBois AB: Vitamin C and airways. Ann NY Acad Sci 1987;498:259–268.
24 Cook NR, Evans DA, Scherr PA, Speizer FE, Taylor JO, Hennekens CH: Peak expiratory flow rate and 5-year mortality in an elderly population. Am J Epidemiol 1991;133:784–794.

25 Taylor A, Jacques PF, Dorey CK: Oxidation and aging: Impact on vision; in Williams GM (ed): Antioxidants: Chemical, Physiological, Nutritional, and Toxicological Aspects. Princeton, Princeton Science Publishers, 1993, pp 349–371.
26 Jacques PF: Antioxidants and cataracts. Epidemiology 1993;191–193.
27 Sperduto RD, Hu T-S, Milton RC, Zhao J-L, Evertt DF, Cheng Q-F, Blot WJ, Bing L, Taylor PR, Jun-Yao L, Dawsey S, Guo W-D: The Linxian cataract studies: Two nutrition intervention trials. Arch Ophthalmol 1993;111:1246–1253.
28 Robertson JM, Donner AP, Trevithick JR: Vitamin E intake and risk of cataract in humans. Ann NY Acad Sci 1989;570:372–382.
29 Eye Disease Case-Control Study Group: Antioxidant status and neovascular age-related macular degeneration. Arch Ophthalmol 1993;111:104–109.
30 West S, Vitale S, Hallfrisch J, Muñoz B, Muller D, Bressler S, Bressler NM: Are antioxidants or supplements protective for age-related macular degeneration? Arch Ophthalmol 1994;112:227.
31 Kaminski MS, Yolton DP, Jordan WT, Yolton RL: Evaluation of dietary antioxidant levels and supplementation with ICAPS-Plus and Ocuvite. J An Optom Assoc 1993;64:862–870.
32 Collier A, Wilson R, Bradley H, Thomson JA, Small M: Free radical activity in type II diabetes. Diabetic Med 1990;7:27–30.
33 Ceriello A, Giugliano D, Quatraro A, Donzella C, Dipalo G, Lefebvre PJ: Vitamin E reduction of protein glycosylation in diabetics: New prospect for prevention of diabetic complications. Diabetes Care 1991;14:68–72.
34 Paolisso G, D'Amore A, Giugliano D, Ceriello A, Varricchio M, D'Onofrio F: Pharmacologic doses of vitamin E improve insulin action in healthy subjects and non-insulin-dependent diabetic patients. Am J Clin Nutr 1993;57:650–656.
35 Paolisso G, D'Amore A, Galzerano D, Balbi V, Giugliano D, Varricchio M, D'Onofrio F: Daily vitamin E supplements improve metabolic control but not insulin secretion in elderly type II diabetic patients. Diabetes Care 1993;16:1433–1437.
36 Paolisso G, Giugliano D, Pizza G, Gambardella A, Tesauro P, Varricchio M, D'Ondofrio F: Glutathione infusion potentiates glucose-induced insulin secretion in aged patients with impaired glucose tolerance. Diabetes Care 1992;15:1–7.
37 Trout DL: Vitamin C and cardiovascular risk factors. Am J Clin Nutr 1991;53:322S–325S.
38 Moran JP, Cohen L, Greene JM, Xu G, Feldman EB, Hames CG, Feldman DS: Plasma ascorbic acid concentrations relate inversely to blood pressure in human subjects. Am J Clin Nutr 1993;57: 213–217.
39 Koh ET: Effect of vitamin C on blood parameters of hypertensive subjects. J Okla State Med Assoc 1984;77:177–182.
40 Parkinson Study Group: DATATOP: a multicenter controlled clinical trial in early Parkinson's disease. Arch Neurol 1989;46:1052–1060.
41 Fahn S: The endogenous toxin hypothesis of the etiology of Parkinson's disease and a pilot trial of high-dosage antioxidants in an attempt to slow the progression of the illness. Ann NY Acad Sci 1989;570:186–196.
42 The Parkinson Study Group: Effects of tocopherol and deprenyl on the progression of disability in early Parkinson's disease. N Engl J Med 1993;328:176–183.
43 Adler LA, Peselow E, Rotrosen J, Duncan E, Lee M, Rosenthal M, Angrist B: Vitamin E treatment of tardive dyskinesia. Am J Psychiatry 1993;150:1405–1407.
44 Lohr JB, Cadet JL, Lohr MA, Larson L, Wasli E, Wade L, Hylton R, Vidoni C, Jeste DV, Wyatt RJ: Vitamin E in the treatment of tardive dyskinesia: The possible involvement of free radical mechanisms. Schizophr Bull 1988;14:291–296.

Dr. Jeffrey B. Blumberg, USDA Human Nutrition Research Center on Aging, Tufts University, 711 Washington Street, Boston, MA 02111 (USA)

Walter P (ed): The Scientific Basis for Vitamin Intake in Human Nutrition.
Bibl Nutr Dieta. Basel, Karger, 1995, No 52, pp 116–127

Vitamin Intake and Vitamin Status in Germany

Roland Schneider[a], *Winfried Eberhardt*[a], *Helmut Heseker*[b], *Werner Kübler*[a]

[a] Institute of Nutritional Sciences, University of Giessen, and
[b] Faculty 06, University-GH-Paderborn, Germany

Before 1987, several studies were conducted in Germany which estimated vitamin intake and vitamin status [for more information, see 1]. However, these studies are based on regional subsamples and therefore did not provide information about vitamin intake and status of the whole population. In 1987 the VERA study (Verbundstudie Ernährungserhebung und Risikofaktoren Analytik[1]) was initiated as the first, and so far only, cross-sectional survey, representative for Germany. The VERA study provides information about tissue levels (nutrients, toxicants), nutrient intake (7-day food records), food patterns (food frequency), health-related parameters, and socioeconomic features in a representative sample of 2,006 adults [2]. This paper represents results of the intake and status of vitamin C, E, B_{12}, carotene, and folate based on the VERA study. Special emphasis will be put on specific subgroups (elderly, smokers, alcohol consumers, obese people). Remarkable findings will be described. Complete results are published elsewhere [1, 3–7].

Methods

The VERA study was carried out on a sample of 2,006 healthy people aged 18–88 years living in private homes in Germany in 1987–1988. This multistaged, stratified random

[1] Nutrition Survey and Risk Factors Analysis. Supported by the German Ministry of Research and Technology, grant 704752, 704754. The author is responsible for the content of this publication.

sample was representative for the noninstitutionalized population in West Germany [8, 9]. The participation rate was 72%. Representativeness of the sample was evaluated by comparing some relevant characteristics of the sample with the total German population [10]. Age, sex, and smoking habits were obtained from a detailed questionnaire on various lifestyle factors. A person was defined as 'smoker' when the respondent reported smoking one or more cigarettes a day. Body weight and height were measured under standardized conditions in order to calculate the body mass index (BMI). Values for daily alcohol and vitamin intake were calculated as arithmetic means from a 7-day food record [5].

Calculating the main sources of vitamin intake in the German diet, all foods were aggregated into 24 food groups. Figures 1, 3 and 5 show the percentages of the total vitamin C, E, and carotene intake for main food groups. Plasma vitamin C was measured photometrically, plasma α-tocopherol and β-carotene by HPLC method, and vitamin B_{12} and folate by RIA method [11]. The blood specimens were collected under fasting conditions. All analyses were carried out under strictly standardized conditions.

For further analyses several subgroups were formed. Stratification considered three age groups (18–35, 36–50, >50 years), two groups for smoking habits (0 and >0 cigarettes/day), three groups for BMI (w ≤24, 24–30, >30; m ≤25, 25–30, >30) and three groups (tertiles) for alcohol consumption (w 0–1.4, 1.4–6, >6; m 0–6, 6–17.1, >17.1 g ethanol/day).

In order to calculate percentages of low vitamin intake and low plasma concentrations in specific subgroups, cut-off points were defined as the 5th percentile of all men and all women, respectively (tables 1, 2). Therefore, the expected percentage in subgroups is 5%. It is of special interest which subgroups have significantly higher or lower percentages than the expectation of 5%. In figures 2, 4 and 6–8, the percentages are shown as bars, the expectation as a horizontal line and the 95% confidence interval is shown as a shaded area. For these sample sizes, the 95% confidence interval lies within the range of 1–9%. There is no significant difference from the expectation ($p>0.05$) if the percentages are within the 95% confidence interval. There is a significant difference from the expectation ($p\leq 0.05$) if the percentage of low values in a specific subgroup is above or below the 95% confidence interval.

Results

The medians (50th percentile) of vitamin intake and plasma concentrations are shown in table 1 (men) and table 2 (women), respectively. Additionally, the 5th and 95th percentile as well as the Recommendations on Nutrient Intake (RNI) of the German Society of Nutrition [12] are shown in these tables.

The median vitamin C intake in Germany is 75.5 mg/day for men and 86.1 mg/day for women, respectively (tables 1, 2). The median plasma vitamin C level is 68.7 µmol/l for men and 81.76 µmol/l for women, respectively. Main sources of vitamin C intake in the German diet are fruit juices and soft drinks, fresh vegetables, and potatoes which provide altogether approximately two thirds of the total vitamin C intake (fig. 1). There is no influence of alcohol

Table 1. Percentiles for plasma concentrations and intake levels (VERA study, males, n = 862)

	Percentiles			RNI[b]
	5[a]	50	95	
Intake				
Vitamin C[c], mg/day	25.20	75.53	220.1	75.0
Vitamin E[c], mg/day	7.60	14.87	29.36	12.0[d]
Carotene[c], mg/day	0.47	1.45	5.53	2.0[e]
Folate[c], μg equiv./day	77.07	138.8	285.2	150.0
Vitamin B_{12}[c], μg/day	3.11	6.25	14.81	3.0
Plasma				
Vitamin C, μmol/l	26.69	68.70	105.6	
α-Tocopherol, μmol/l	18.77	29.15	47.59	
β-Carotene, μmol/l	0.12	0.43	1.29	
Folate, nmol/l	7.00	13.00	23.00	
Vitamin B_{12}, pmol/l	137.0	276.0	525.0	

[a] Cut-off points for intake and plasma levels.
[b] Recommendations on Nutrient Intake of the German Society of Nutrition (DGE, 1991).
[c] Seven-day-records.
[d] Tocopherol equivalents.
[e] Guideline value for β-carotene.

consumption on vitamin C intake and status (data not shown). On the other hand, smoking habits influence vitamin C status very strongly. Smokers show lower vitamin C intake and lower vitamin C plasma concentrations, the latter even when controlled for vitamin C intake [7].

Assessing the influence of age, stratification for smoking habits was necessary. The following results represent smokers only. In both sexes, there are higher percentages of low plasma vitamin C concentrations among elderly smokers compared to younger smokers (fig. 2). The percentages are significantly higher than the expectation and demonstrate clearly the worse vitamin C status among the elderly. In contrast to these findings, percentages of low vitamin C intake among the elderly comply with the expected value. Among younger women, vitamin C intake is relatively low (significantly higher percentages compared to the expected value). Tests showed that differences in smoking habits between age groups are not responsible for these results. The proportion of heavy smokers (>21 cigarettes/day) in the age group >50 years is 22.5%, in other age groups >40%.

Table 2. Percentiles for plasma concentrations and intake levels (VERA study, females, n = 1,144)

	Percentiles			RNI[b]
	5[a]	50	95	
Intake				
Vitamin C[c], mg/day	26.32	86.13	235.1	75.0
Vitamin E[c], mg/day	6.38	12.49	24.27	12.0[d]
Carotene[c], mg/day	0.49	1.64	6.25	2.0[e]
Folate[c], μg equiv./day	58.63	117.8	249.8	150.0
Vitamin B_{12}[c], μg/day	1.97	4.39	11.91	3.0
Plasma				
Vitamin C, μmol/l	41.45	81.76	114.7	
α-Tocopherol, μmol/l	19.12	29.15	47.16	
β-Carotene, μmol/l	0.21	0.61	1.77	
Folate, nmol/l	7.00	13.00	26.00	
Vitamin B_{12}, pmol/l	143.0	274.0	555.0	

[a] Cut-off points for intake and plasma levels.
[b] Recommendations on Nutrient Intake of the German Society of Nutrition (DGE, 1991).
[c] Seven-day-records.
[d] Tocopherol equivalents.
[e] Guideline value for β-carotene.

The median vitamin E intake in Germany is 14.9 mg/day for men and 12.5 mg/day for women, respectively (tables 1, 2). The median plasma α-tocopherol level is 29.2 μmol/l for men and women. Main sources of vitamin E intake in the German diet are vegetable oils/fat, bread/cakes, and grain products/cereals which provide altogether approximately two thirds of the total vitamin E intake (fig. 3).

Alcohol consumption as well as smoking influence vitamin E intake and status [13]. Smokers and nonsmokers with higher alcohol consumption show lower vitamin E intake and lower plasma α-tocopherol concentrations (significantly higher percentages compared with the expectation).

In both genders, there are lower percentages of low plasma α-tocopherol concentrations among elderly smokers compared with younger smokers (fig. 4). However, there is no influence of age on plasma α-tocopherol concentrations if plasma lipoproteins are considered as confounding factor [14]. On the other hand, the percentages of low vitamin E intake among elderly men and women is significantly higher than the expectation.

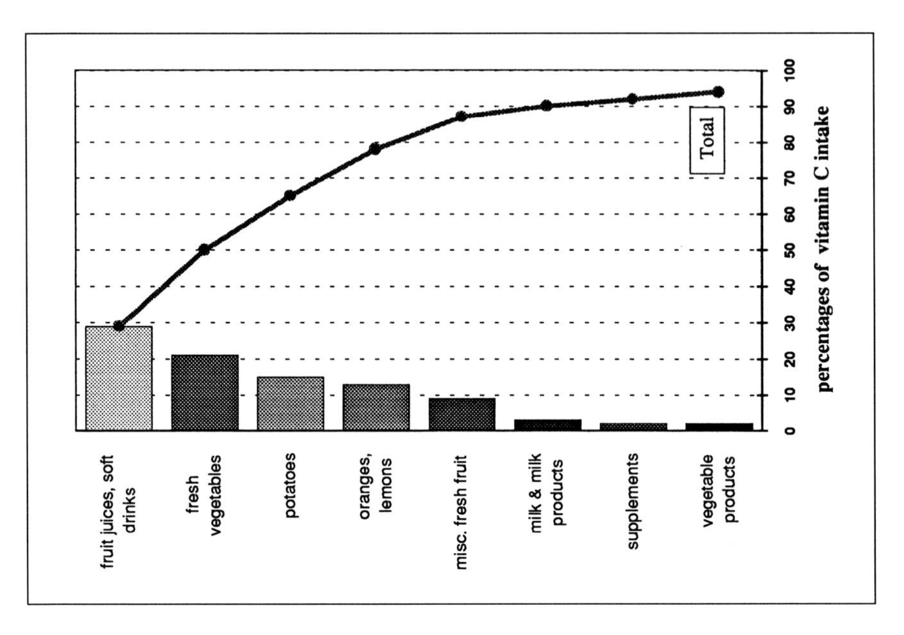

Fig. 1. Main sources of vitamin C in the German diet (VERA, 1987/88, n = 2,006).

The median carotene intake in Germany is 1.45 mg/day for men and 1.64 mg/day for women, respectively (tables 1, 2). the median plasma β-carotene level is 0.43 μmol/l for men and 0.61 μmol/l for women, respectively. It should be mentioned that the distributions for carotene intake and plasma β-carotene are extremely skewed. Especially the 95th percentile of carotene intake is extremely high and the arithmetic mean is much higher than the median (e.g. male: median 1.45 mg/day, arithmetic mean 2.0 mg/day). The main sources of carotene intake in the German diet are fresh vegetables and fruit juices/soft drinks which provide altogether more than two thirds of the total carotene intake (fig. 5). There are three important factors which influence carotene intake and plasma levels: smoking, alcohol consumption, and body weight. As already shown, nonsmokers have higher plasma β-carotene levels than smokers [1]. It should also be mentioned that alcohol consumption, independently from cigarette consumption, influences the supply with β-carotene [7]. The following results will focus on the influence of relative body weight (BMI) for nonsmokers only. Similar results were observed for smokers. Further stratification for alcohol consumption was not possible because of the relatively small sample size. Figure 6 shows clearly that the higher the BMI, the higher percentages of low plasma β-carotene levels and low carotene intake values

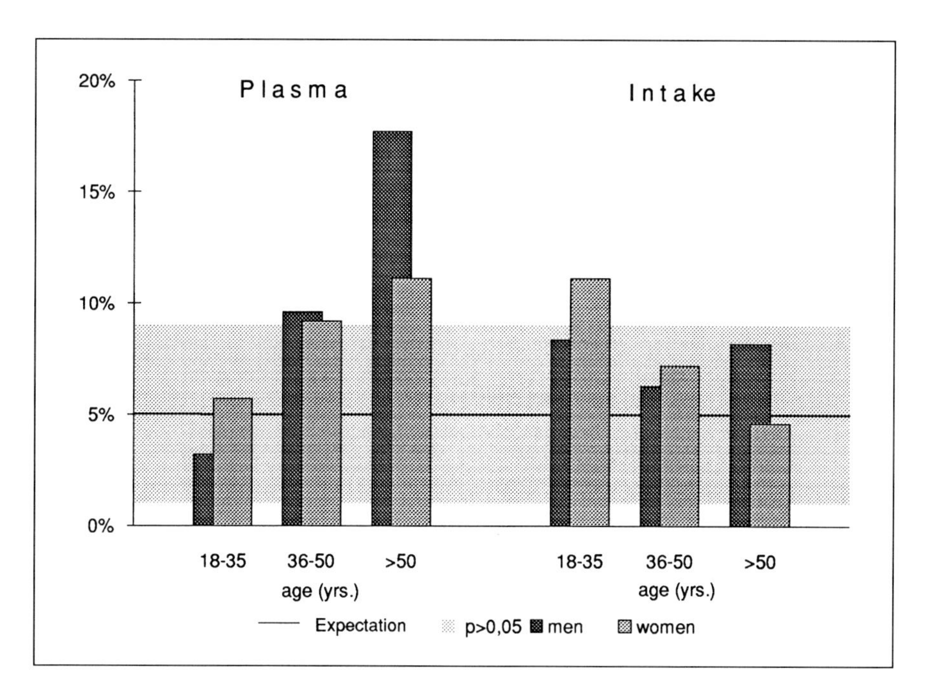

Fig. 2. Frequencies below the cut-off point for vitamin C depending on age (VERA, 1987/88, smokers only, n = 821).

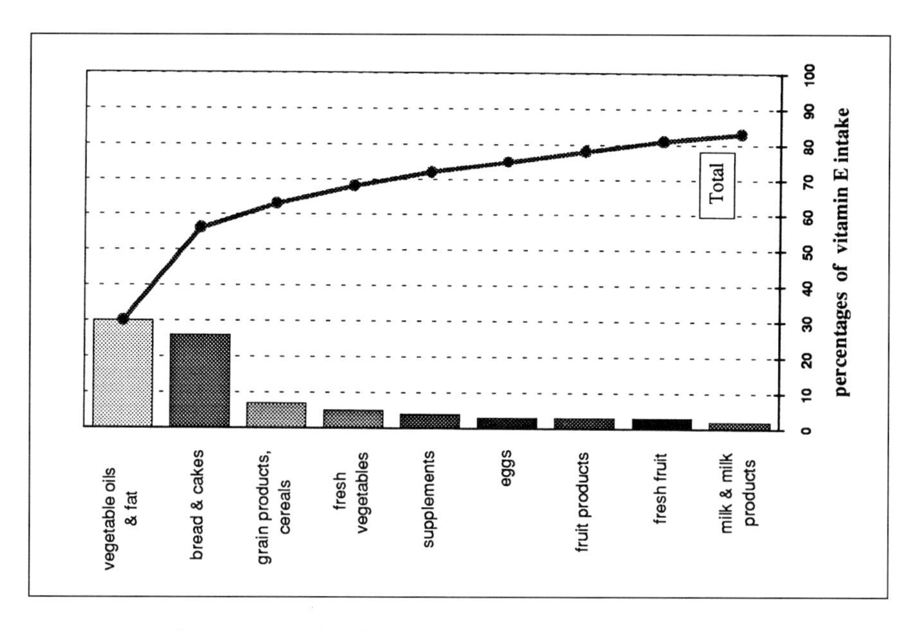

Fig. 3. Main sources of vitamin E in the German diet (VERA, 1987/88, n = 2,006).

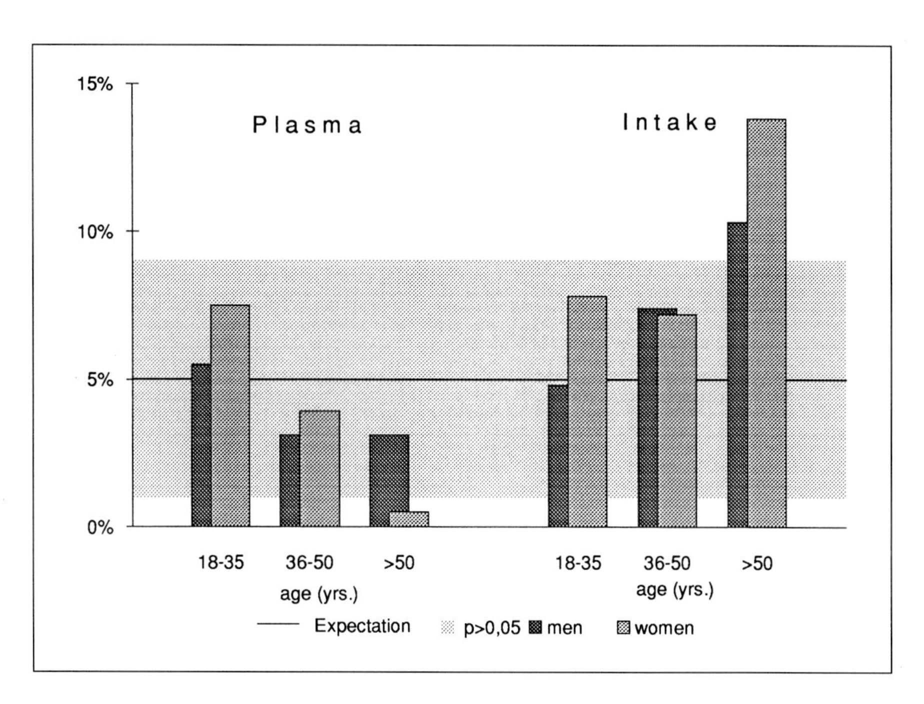

Fig. 4. Frequencies below the cut-off point for vitamin E depending on age (VERA, 1987/88, smokers only, n = 821).

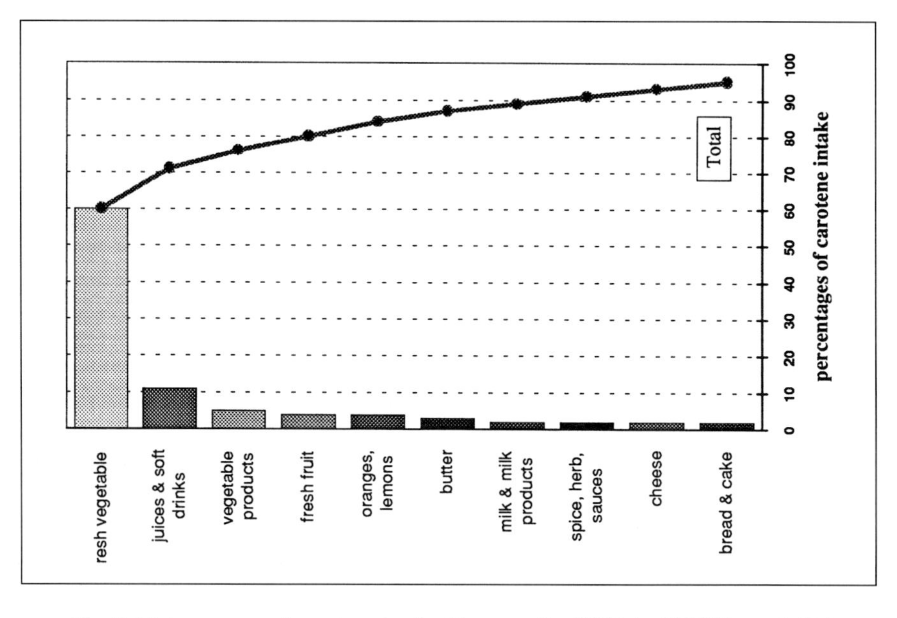

Fig. 5. Main sources of carotene in the German diet (VERA, 1987/88, n = 2,006).

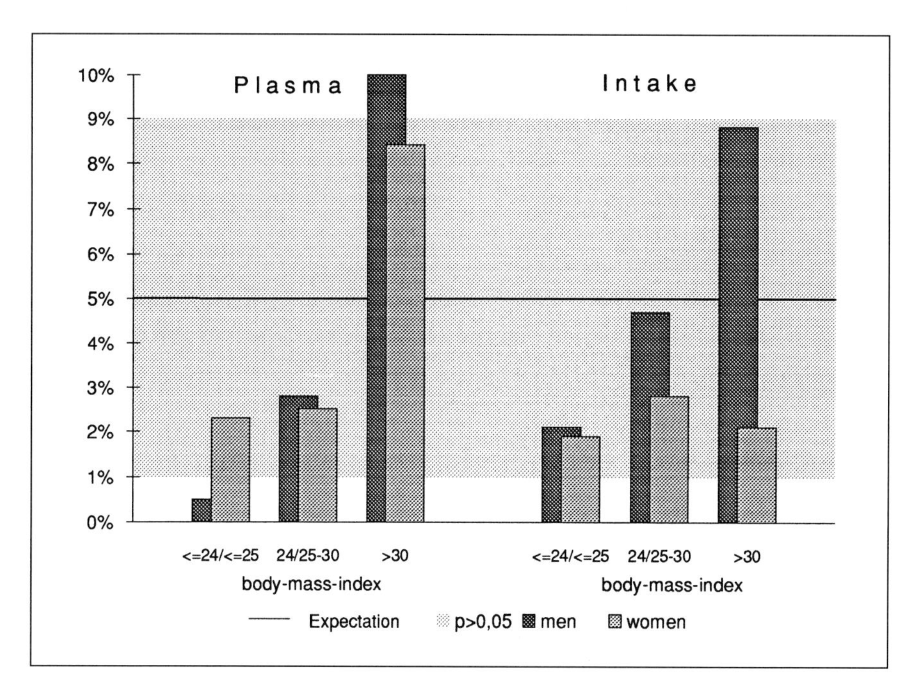

Fig. 6. Frequencies below the cut-off point for carotene depending on BMI (VERA, 1987/88, nonsmokers only, n = 1,185).

will be observed. For men the percentages of low plasma β-carotene concentrations are significantly different from expectation in the lowest and in the highest BMI group. there were no significant differences in plasma lipoproteins between the different groups.

The median folate intake in Germany is 138.8 μg equivalents/day for men and 117.8 μg equivalents/day for women, respectively. The median plasma folate level is 13.0 nmol/l for men and women (tables 1, 2). The influence of alcohol consumption on folate is shown in figure 7. Alcohol consumption is classified in tertiles, so that similar sample sizes will be obtained. Results will be shown for men only.

Highest percentages below the cut-off point can be seen in the lowest alcohol intake group for intake and plasma levels of folate. The percentages are significantly different from expectation, expecially for smokers. They show higher percentages of low values than nonsmokers.

The median vitamin B_{12} intake in Germany is 6.25 μg/day for men and 4.39 μg/day for women, respectively (tables 1, 2). The median plasma vitamin B_{12} level is 276 pmol/l for men and 274 pmol/l for women, respectively. The

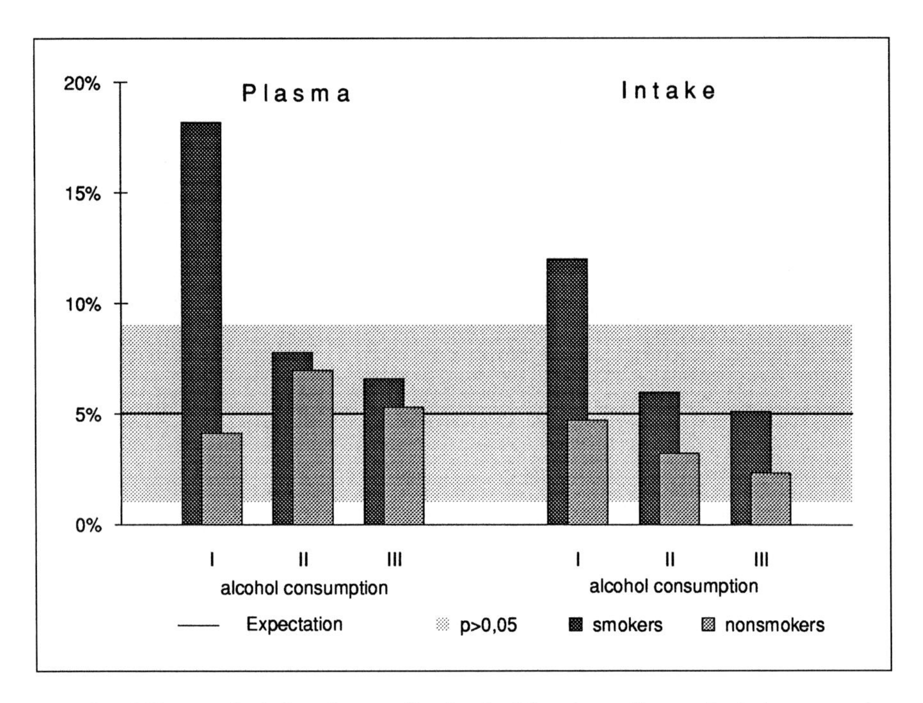

Fig. 7. Frequencies below the cut-off point for folate depending on alcohol consumption (VERA, 1987/88, men, n = 862).

influence of alcohol consumption on vitamin B_{12} is shown in figure 8. Similiar to folate, the prevalences of low vitamin B_{12} intake are highest for low alcohol consumption and lowest for high alcohol consumption, indicating that consumption of alcoholic beverages provides higher intake of vitamin B_{12}. For nonsmokers the percentage is significantly different from expectation. In contrast to plasma folate, plasma vitamin B_{12} levels do not differ depending on alcohol consumption.

Discussion

Data of the representative cross-sectional VERA study were used to describe plasma and intake levels of vitamin C, Vitamin E, carotene, vitamin B_{12}, and folate. Besides, remarkable findings of differences in subgroups of the German population (elderly, smokers, alcohol consumers, obese people) were described. Interpretation of results from cross-sectional surveys is sometimes difficult because many confounding factors have to be taken into account. For

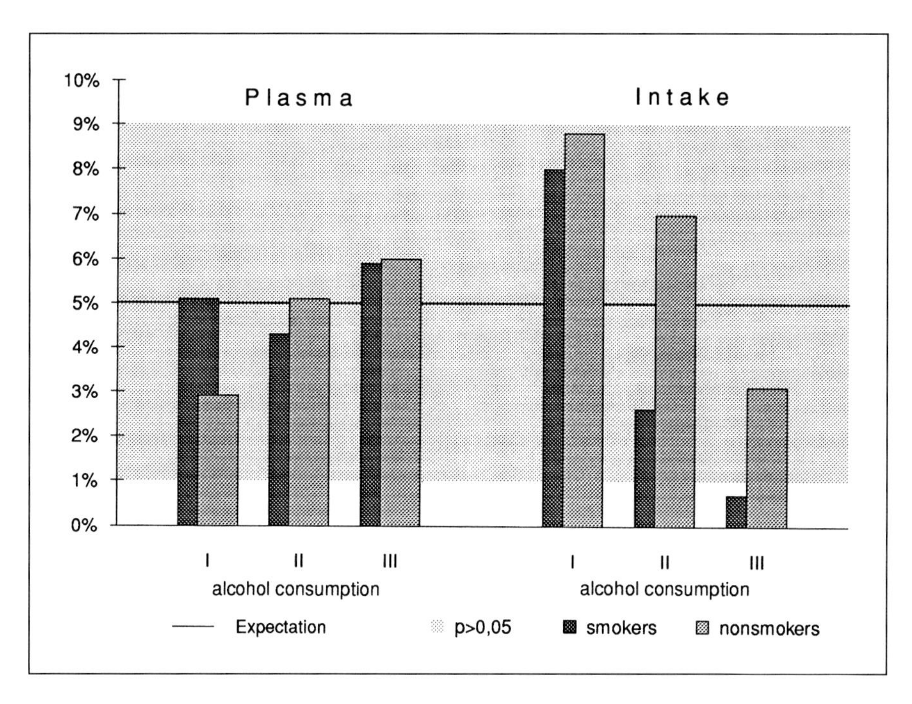

Fig. 8. Frequencies below the cut-off point for vitamin B_{12} depending on alcohol consumption (VERA, 1987/88, men, n = 862).

example, smoking and alcohol consumption independently influence plasma β-carotene levels [7]. On the other hand, the absolute characterization of intake or plasma levels as adequate or inadequate can be very misleading. Reasons for this are the lack of data about nutrient requirements of humans, insufficient comparability of laboratory assessment methods and dietary assessment methods, large variability in parameters, and insufficient data in nutrient databases [15, 16]. This leads to the problem that there are no or only inadequate data for the implementation of absolute cut-off points. Similarly, it does not seem appropriate to use the RNI values as cut-off points for vitamin intake because of insufficient information about requirements and problems of underreporting in food records. For these reasons, this paper used statistically defined cut-off points. The cut-offs were defined as the 5th percentile of the whole sample so that there is an expectation in subgroups of 5% below these cut-offs. The actual percentages in subgroups can be compared to the expectation. The expectation of 5% is valid for all plasma and intake parameters.

There are at least three advantages of calculating percentages of low values and comparing these percentages with the expectation. First, it is possible to focus on subjects with a low intake and low plasma levels. Second, it is possible to compare the results of vitamin intake and vitamin plasma levels with the same expectation. Third, the definition of the cut-off as the 5th percentile of the whole group is a clear and understandable concept which does not represent more than is intended. That means, subjects who are below the cut-off point do not necessarily have a vitamin deficiency in any way. The concept was developed exclusively to compare intake and plasma levels of subgroups.

For all researched vitamins there is a strong influence of smoking. Smokers almost always have higher percentages of values below the cut-off points. Similar results were found for the elderly, who often show higher percentages of low values than younger volunteers (e.g. vitamin E intake).

In some cases a paradoxical observation was made. Although vitamin intake was comparably high, plasma levels were lower compared with younger age groups. This is true for vitamin C where elderly smokers show lower levels although vitamin C intake is comparably higher. One explanation could be that elderly smokers have a higher requirement for vitamin C. Even with a higher intake, elderly smokers cannot achieve plasma vitamin C levels comparable with younger smokers. This hypothesis is supported by further analyses of the same data [3].

Alcohol consumption also influences vitamin intake and status. In contrast to smoking or age, high consumption of alcohol leads to low percentages of low values for folate and vitamin B_{12} intake. It seems that higher intake of alcoholic beverages provides higher folate and vitamin B_{12} intake. This might be due to the high intake of beer in Germany which is rich in folate and vitamin B_{12}. The higher intake of folate also results in lower percentages of low plasma folate levels for subjects with higher alcohol intake compared to subjects with lower alcohol intake. In contrast to plasma folate, alcohol consumption has no effect on percentages of low plasma vitamin B_{12}. The observation that higher alcohol consumption results in higher vitamin B_{12} intake but not in higher plasma levels is caused probably by insufficient resorption in the small intestine. This hypothesis is supported by the fact that the process of vitamin B_{12} resorption depends on the intrinsic factor which is produced in the stomach and, on the other hand, will be affected by higher alcohol consumption. Similarly, negative effects of alcohol consumption on plasma concentrations can be observed for vitamin E and carotene.

According to the vitamin intake and vitamin status, people should stop smoking, reduce consumption of alcohol other than beer, and stop getting older.

References

1 Schneider R: Die Beurteilung der Nährstoffversorgung bundesdeutscher Bevölkerungsgruppen am Beispiel von Vitamin C und β-Carotin (Auswertungskonzepte). VERA-Schriftenreihe, vol VIII. Niederkleen, Dr-Fleck-Verlag, 1992, p 189.

2 Kübler W, Hüppe R, Matiaske B, Rosenbauer J, Anders HJ: Was verzehrt der Bundesbürger? – Was sind die Folgen? Ernähr Umsch 1990;37:102–107.

3 Heseker H, Schneider R: Requirement and supply of vitamin C, E, and β-carotene for elderly men and women. Eur J Clin Nutr 1994;48:118–127.

4 Heseker H, Schneider R, Moch KJ, Kohlmeier M, Kübler W: Vitaminversorgung Erwachsener in der Bundesrepublik Deutschland. VERA-Schriftenreihe, vol IV. Niederkleen, Dr-Fleck-Verlag, 1992, pp 26, 85, 94.

5 Heseker H, Adolf T, Eberhardt W, Hartmann S, Herwig A, Kübler W, Matiaske B, Moch KJ, Schneider R, Zipp A: Lebensmittel- und Nährstoffaufnahme Erwachsener in der Bundesrepublik Deutschland. VERA-Schriftenreihe, vol III. Niederkleen, Dr-Fleck-Verlag, 1992, p 9.

6 Schneider R: Wer ist unzureichend versorgt? Auswertungskonzepte zur Identifizierung von Risikogruppen. Ernähr Umsch 1993;12:480–485.

7 Schneider R: Where are the gaps in assessing the relationship between nutritional status parameters and nutrient intake data? Bibl Nutr Diet. Basel, Karger, 1995, No 51, pp 74–83.

8 Anders HJ, Rosenbauer J, Matiaske B: Repräsentative Verzehrsstudie in der Bundesrepublik Deutschland inkl. West-Berlin. Schriftenr AGEV, vol 8. Frankfurt, Umschau Verlag, 1990.

9 Schaefer F: Muster-Stichproben-Pläne für Bevölkerungsstichproben in der Bundesrepublik Deutschland und West-Berlin; in Arbeitskreis deutscher Marktforschungsinstitute (ADM). München, Verlag Moderne Industrie, 1979.

10 Schneider R, Eberhardt W, Heseker H, Moch KJ: Die VERA-Stichprobe im Vergleich mit Mikrozensus, Volkszählung und anderen nationalen Untersuchungen. VERA-Schriftenreihe, vol II. Niederkleen, Dr-Fleck-Verlag, 1992, p 17.

11 Speitling A, Kohlmeier M, Matiaske B, Stelte W, Thefeld W, Wetzel S: Methodological Handbook: Nutrition Survey and Risk Factors Analysis. VERA-Schriftenreihe, vol Ia. Niederkleen, Dr-Fleck-Verlag, 1992, pp 74, 102.

12 Deutsche Gesellschaft für Ernährung (ed): Empfehlungen für die Nährstoffzufuhr, ed 5. Frankfurt, Umschau, 1991.

13 Schneider R, Eberhardt W, Heseker H, Kübler W: Vitamine von A (wie Alkohol) bis Z (wie Zigaretten) – Die Vitaminversorgung in Deutschland im Spiegel des Genussmittelkonsums unter Berücksichtigung des BAVE-Konzeptes. Ernährung/Nutrition 1994;18:349–354.

14 Heseker H, Kohlmeier M, Schneider R: Lipid-Adjustierung von α-Tocopherolkonzentrationen im Plasma. Z Ernährungswiss 1993;32:219–228.

15 Beecher G, Matthews RH: Nutrient composition of foods; in Myrtle, Brown (ed): Present Knowledge in Nutrition. Washington, ILSI, 1990, pp 430–433.

16 Pao EM, Cypel YS: Estimation of dietary intake; in Myrtle, Brown (ed): Present Knowledge in Nutrition. Washington ILSI, 1990, pp 399–406.

17 Kübler W: Wie kann die Bedarfsdeckung mit essentiellen Nährstoffen beurteilt werden? Problemanalyse am Beispiel von jungen Erwachsenen und Senioren. Ernährung/Nutrition 1986;10:83–87.

Dr. Roland Schneider, Infratest Epidemiology and Health Research,
Landsberger Straße 338, D–80687 Munich (Germany)

Walter P (ed): The Scientific Basis for Vitamin Intake in Human Nutrition.
Bibl Nutr Dieta. Basel, Karger, 1995, No 52, pp 128–136

Vitamin Intake in Great Britain: Association with Mortality Rates for Coronary Heart Disease

Margaret Ashwell[a], *David Buss*[b]

[a] British Nutrition Foundation, London, and
[b] Ministry of Agriculture Fisheries and Food, London, UK

This paper will focus on intakes of the antioxidant nutrients, vitamin C, vitamin E and carotenes in Britain although these three nutrients are only part of the antioxidant potential of the diet. A special emphasis will be given to the social class and regional variations in the intakes of these vitamins because of the substantial differences in standardized mortality rates (SMRs) for coronary heart disease (CHD) that are seen between different social classes and across the different geographical regions within Britain [1] which are not explained by classical risk factors [2].

Vitamin Intakes from Household Dietary Surveys

There are three levels at which vitamin intakes can be determined in most countries, including Britain:

The *first* is from estimates of total food supplies, as given in FAO 'Food Balance Sheets', but although the methodology should be comparable between almost all countries of the world, such estimates are somewhat unhelpful and will not be considered further.

The *second* is from surveys of household diets. The amounts and types of foods recorded in these are much closer to what is actually eaten, and Britain's National Food Survey [3] is a well-known example. It has been running continuously since 1940, and each year shows the amounts of vitamin C and carotenes in Britain as a whole as well as in subgroups such as Scotland, Wales and the seven main regions of England, and households with different

Table 1. Household intakes of antioxidant nutrients, 1992 (National Food Survey) [source: 3]

	Vitamin C mg/day	Carotenes μg/day	Vitamin E mg/day
Average	51	1,750	8.1
England	51	1,750	8.1
Scotland	50	1,670	7.8
Wales	48	1,870	7.7
Income group A (highest)	62	1,830	8.1
Income group B	52	1,750	8.1
Income group C	46	1,690	7.8
Income group D (lowest)	44	1,570	8.3

incomes and family composition. Vitamin E was recently added to the nutrients evaluated, and selected results for 1992 are shown in table 1.

There are only small regional and income differences in vitamin E intake within Britain, and intakes of β-carotene are very dependent on the popularity of carrots, which provide more than half the total carotene intake. There are, however, larger variations in vitamin C because of the marked regional and income differences in fruit and green vegetable consumption in Britain. But although intakes of vitamin C have long been lower in Scotland and northern England than in the south, and in poorer households than in richer ones, the regional differences in particular have become smaller in recent years as fruit juices (which are widely consumed) have become an important source. Seasonal differences have also become smaller now that fresh fruit and vegetables (and frozen vegetables) are available throughout the year. Parenthetically, though, it is worth mentioning that the concentration of all three nutrients in foods can vary very widely (e.g. the amount of β-carotene in carrots was 12,000 μg/100 g in 1978 British Food Tables [4] but only 8,115 μg in the 1991 Tables [5]. It has also been quoted anywhere from 4,000 to 25,000 μg in other studies). This variability should be remembered when comparing intakes between countries.

Vitamin Intakes from Dietary Surveys of Individuals

The *third* way of estimating antioxidant nutrient intakes is from dietary surveys of individuals. Britain again has a number of excellent surveys of this type, and some of the results from the dietary and nutritional survey of

Table 2. Intakes of antioxidant nutrients by adults, 1986/87 [source: 2]

	Vitamin C mg/day	Carotenes μg/day	Vitamin E mg/day
Men (food only)	66	2,410	9.9
Men (incl. pills)	75		11.7
Women (food only)	62	2,129	7.2
Women (incl. pills)	73		8.6

some 2,200 adults conducted across the whole of Britain in 1986/87 [2] are shown in table 2 (The Adult Survey). The amounts of all three nutrients are somewhat higher than in the National Food Survey, partly because the Adult Survey included meals outside the home, but also because it covered only adults, and children are known to eat less food and therefore, in general, less of all the nutrients. Indeed, intakes per MJ are very similar in each survey.

Table 2 brings out the significant contribution that dietary supplements made in 1987. Supplement usage was greatest among higher income and southern women (who tended to have high intakes even without the supplements), and supplement usage has become even more widespread since then.

Figure 1 shows the distribution of intakes of the three antioxidant nutrients for adults not consuming pills [6].

Figure 2 shows the variation in intake for these nutrients according to social class. As in the results from the National Food Survey [3], there are substantial social class differences for vitamin C and for carotenes in both men and women. Adults in the higher social classes (I and II) tended to have intakes above the national average and those in lower social classes (IV and V) tended to have intakes below the national average.

Figure 3 shows the variation in vitamin intake according to the geographical region of residence. Four broad geographical regions were distinguished: Scotland; Northern England; Central England, S.W. England and Wales; and London and the South East. Intakes of all three antioxidant nutrients tended to be less than the national average in Scotland (men) and in the North (women) and tended to be higher than the national average in London and the South East for both men and women.

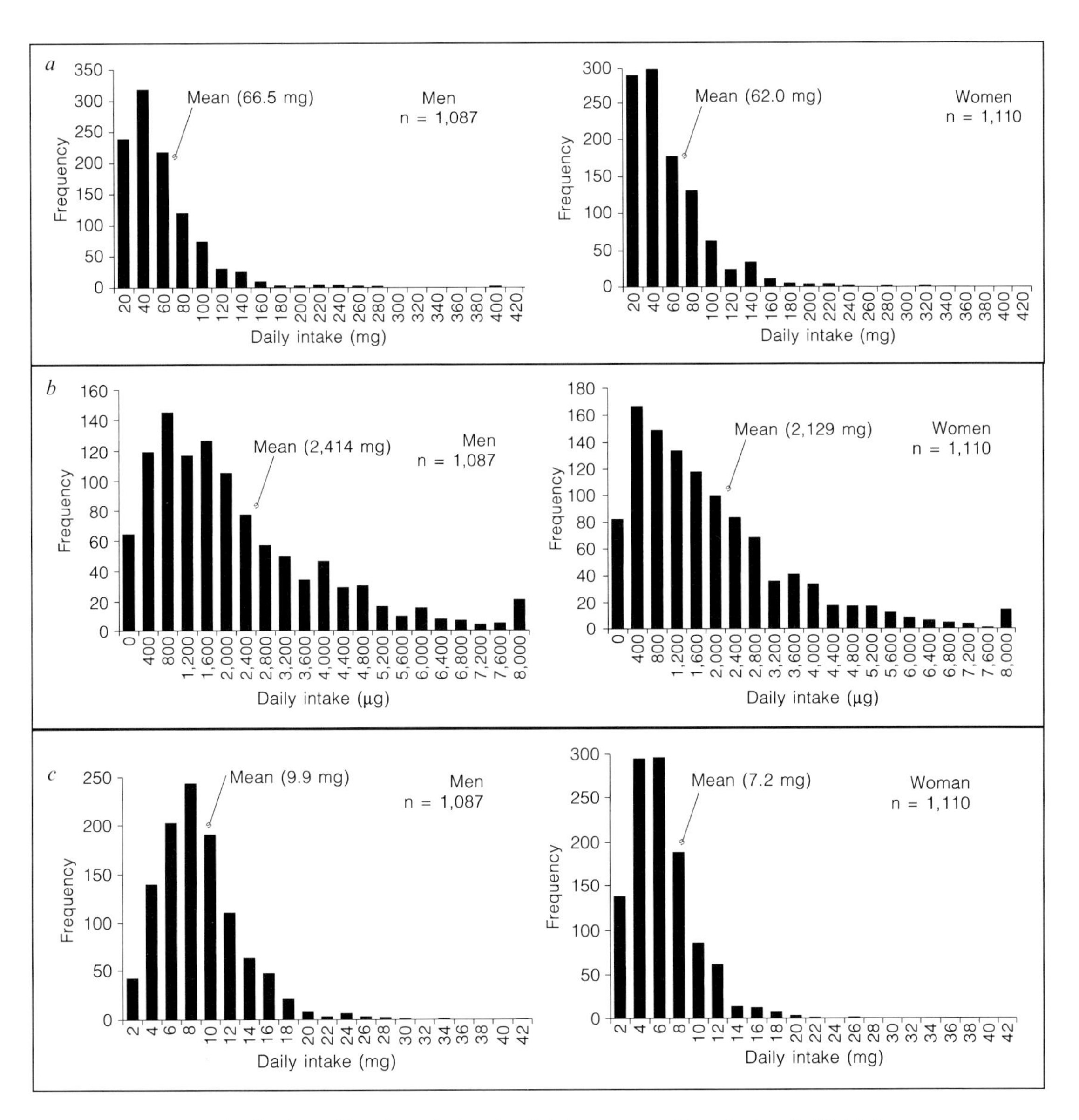

Fig. 1. Distribution of intakes of vitamin C (*a*) carotenes (*b*) and vitamin E (*c*) for British men and women in 1986/7 [2, 6].

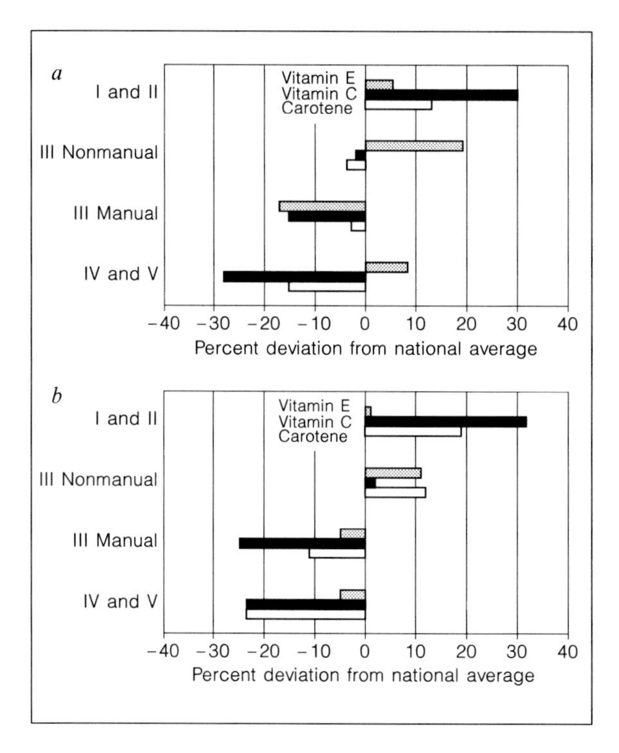

Fig. 2. Variation in vitamin intake by social class for British men (*a*) and women (*b*) in 1986/7 [2, 6].

Other Antioxidants in the Diet

It should also be remembered that vitamin C, β-carotene and vitamin E are by no means the only antioxidants in the diet. Their intakes (50–60, 2, and < 10 mg respectively) need to be compared with the likely intake of other antioxidants such including the nonnutrient carotenoids lycopene and lutein, flavonoids, and the synthetic antioxidants. Average intake of lycopene will not be as high as in Mediterranean countries: British tomato consumption is only around 20 g/day so the average intake is only about 0.2 mg. On the other hand, flavonoids quercetin, kaempferol, myricetin, apigenin and luteolin may be around 25 mg/day as in the Netherlands [7], and intakes of the synthetic antioxidants BHA, BHT and gallates could be as high if they are used to the levels permitted in the forthcoming EC Directive on Food Additives other than Sweeteners.

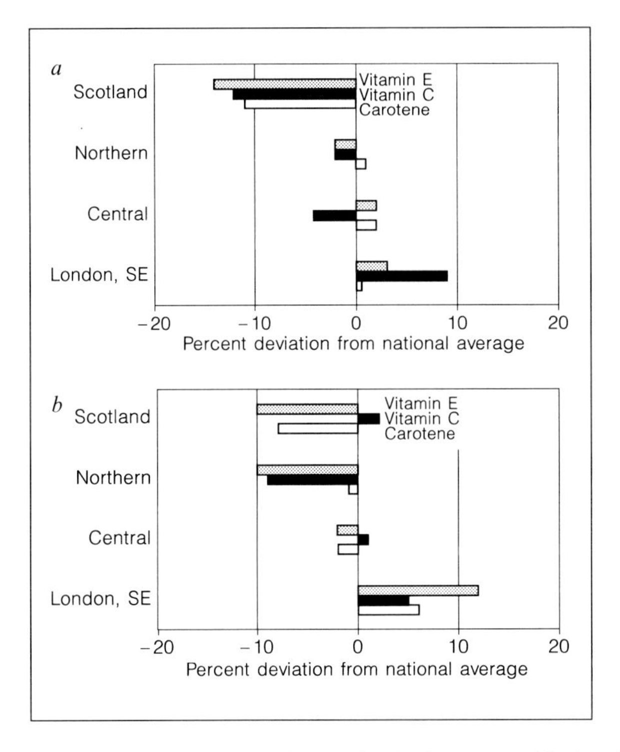

Fig. 3. Variation in vitamin intake by geographical region for British men (*a*) and women (*b*) in 1986/7 [2, 6].

Plasma Levels of Vitamins and Other Antioxidants

Most participants in the Adult Survey, aged 18 and over, gave a blood sample (946 men and 946 women). Among other things, these were analysed for plasma levels of vitamin E, α-carotene, β-carotene and β-cryptoxanthin. The last three values were added together to give an average plasma value for carotene levels.

Mean plasma levels for vitamin E were 27.1 and 26.2 μmol/l for men and women respectively. Plasma vitamin E showed significant positive correlations with dietary intakes of vitamin E for both men and women ($p < 0.01$).

Mean plasma levels for carotenes were 0.53 and 0.68 μmol/l for men and women respectively. Plasma levels for carotenes showed significant positive correlations with dietary intake ($p < 0.01$).

Figures 4 and 5 show the social class and regional variations in plasma levels of carotenes and vitamin E. As with dietary intakes of these vitamins, there were distinctive social class and regional gradients which were greater

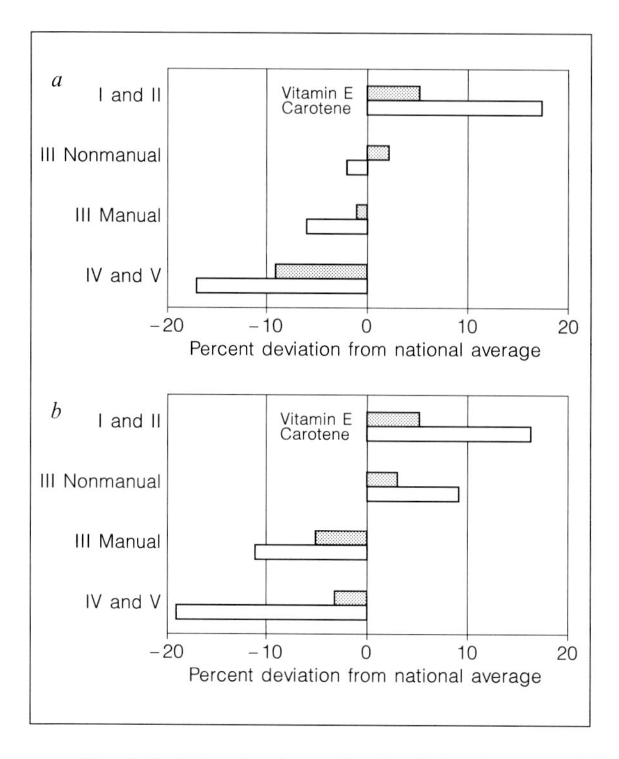

Fig. 4. Variation in plasma levels of vitamins by social class for British men (*a*) and women (*b*) in 1986/7 [2].

for the carotenes than for vitamin E. Mean values for β-carotene in men, for example, ranged from 0.33 μmol/l for those in social classes I and II to 0.24 μmol/l in social classes IV and V ($p<0.01$). Among men living in London and the South East, levels of β-carotene ($p<0.01$) and vitamin E ($p<0.05$) were significantly higher than those for men living in the Northern region.

Plasma levels of lycopene were also measured in the Adult Survey and showed significant social class and regional variations which paralleled those shown by the carotenes. Mean levels for men and women were 0.30 and 0.28 μmol/l respectively. Levels were as low as 0.23 μmol/l for the lowest social class men (IV and V) and as high as 0.34 μmol/l for men in the highest social class (I and II).

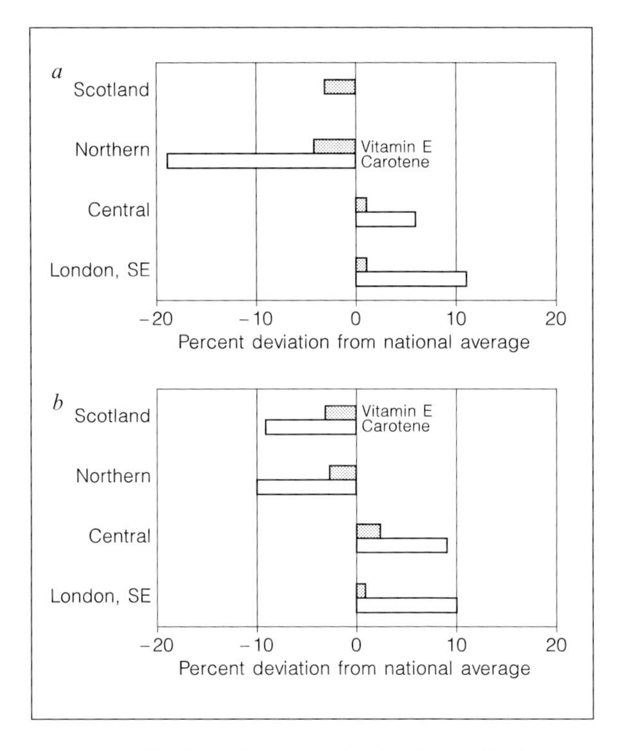

Fig. 5. Variation in plasma levels of vitamins by geographical region for British men (*a*) and women (*b*) in 1986/7. NB: Men in Scotland had carotene levels equal to the national average [2].

CHD Mortality in the UK

Apart from the universally acknowledged age and gender differences in CHD mortality, substantial social class and regional differences can be demonstrated within Britain. SMRs for social class I men and women are substantially below the average (69 and 41 respectively) and SMRs for social class V men and women are substantially above the average (137 and 152 respectively) [8]. Regional differences are also apparent with SMRs in Scotland being substantially higher (up to about 140) than those in London and the South East (below about 85). Although there are more smokers in the lower social classes, this is the only classical risk factor that is associated with the observed regional and social class differences; blood pressure, total cholesterol and HDL cholesterol do not show any significant differences.

Comment

Differences in intake of the antioxidant nutrients, vitamin C, vitamin E and the carotenes, and the plasma levels of vitamin E and carotenes are associated with the observed regional and social class differences in CHD mortality. This observation is particularly intriguing since a similar analysis of intakes of total fat and saturated fatty acids did not reveal any regional or social class differences [2].

The potential role of antioxidants in preventing CHD and cancer will be documented elsewhere in this book. Unfortunately the social class and regional variations for mortality from different cancers in Britain [9] are not as clear cut as the CHD mortality statistics and so it is not possible to make any association between the intake of vitamins discussed here and cancer.

The data presented here for vitamin intake in Britain give some support to the hypothesis that long-term habitual intake of a diet high in antioxidant substances might give some protection against CHD.

References

1 Ashwell M (ed): Diet and Heart Disease: A Round Table of Potential Factors. London, British Nutrition Foundation, 1993.
2 Gregory J, Foster K, Tyler H, Wiseman M: The Dietary and Nutritional Survey of British Adults. London, HMSO, 1990.
3 Ministry of Agriculture Fisheries and Food: National Food Survey 1992. London, HMSO, 1993.
4 Paul AA, Southgate DAT: McCance and Widdowson's The Composition of Foods, ed 4. London, MAFF & MRC, 1978.
5 Holland B, Welch AA, Unwin ID, Buss DH, Paul AA, Southgate DAT: McCance and Widdowson's The Composition of Foods, ed 5. Cambridge, Royal Society of Chemistry, 1991.
6 Ministry of Agriculture Fisheries and Food: The Dietary and Nutritional Survey of British Adults – Further Analysis. London, HMSO, 1994.
7 Hertog MGL, Feskens EJM, Hollman PCH, Katan MB, Kromhout D: Dietary antioxidant flavonoids and risk of coronary heart disease: The Zutphen Elderly Study. Lancet 1993;342:1007–1011.
8 Office of Population Censuses & Surveys. Occupational Mortality: The Registrar-General's Decennial Supplement for Great Britain, 1979–80, series P5, No 6. London, HMSO, 1986.
9 Office of Population Censuses & Surveys. Cancer Statistics Registrations 1987. London, HMSO, 1993.

Dr. Margaret Ashwell, Ashwell Associates, Ashwell Street,
Ashwell, N. Hertfordshire SG7 5PZ (UK)

Walter P (ed): The Scientific Basis for Vitamin Intake in Human Nutrition.
Bibl Nutr Dieta. Basel, Karger, 1995, No 52, pp 137–145

Vitamin Intake in Sweden and Other Nordic Countries

Wulf Becker

National Food Administration, Uppsala, Sweden

Dietary surveys of population groups in Sweden and other Nordic countries have been carried out since the end of the last century. Most of these studies have included specific population groups limited to certain age and sex groups and covering certain geographical regions. National dietary surveys have been carried out in some countries during the last decade. In Sweden, the first nationwide dietary survey, in which a representative sample of the Swedish population participated, was carried out in 1989. Few and limited studies have been made in which the biochemical status of e.g. vitamins were assessed. In this paper, data on intakes of vitamin A, β-carotene, vitamin E, ascorbic acid and folacin from the Swedish national survey are presented as well as intake data from national surveys carried out in some other Nordic countries.

The Swedish HULK Survey

In 1989, a combined nationwide dietary and household budget survey (HBS) was carried out by the Statistics Sweden in cooperation with the National Food Administration (NFA) to obtain quantitative data on the food and nutrient intake among a representative sample of the Swedish population. The survey was based on a representative sample of 3,000 persons aged 0–74 years. The household to which the selected person belonged was included in the HBS. The selected households were divided into 13 consecutive subsamples distributed over the year.

The survey consisted of three parts: (1) an interview covering background information; (2) 4-week record of the households' food purchases (expenditures and amounts), and (3) menu book, a simplified 7-day record with preprinted alternatives for food consumed by the selected household member (excluding children <1 year old).

The menu book gives preprinted alternatives (with quantity indications in household measures) for foods, meal components and facilities for indication of where and when the meals are consumed. Using a portion guide with photographs, sizes of cooked food portions eaten at main meals could be estimated. The use of fat spreads on sandwiches was estimated with the help of an illustration shown in the introductory interview. Snacks and other in-between meal eating were recorded in household measures, number, etc., in the traditional way. About 2,000 persons completed the study with a participation rate of 70%. Participation was lower in larger cities and surrounding areas than in rural areas and lower among young as well as older households.

In table 1 the average intakes of certain vitamins in different age and sex groups are presented and compared with the current Nordic (and Swedish) recommended daily intakes (RI) [2]. In figures 1–5 the distribution of the daily intakes of men and women 15–74 years old are shown. Probably the true daily intakes shown in the figures are somewhat higher since ca. 20% of the participants in this age group were found to have energy intakes that were below the estimated reasonable cut-off limit for energy intake, indicating under-reporting.

Vitamin A

The intake of vitamin A (preformed retinol and vitamin A activity from β-carotene) is generally high in Sweden, partly due to fortification of margarines and low-fat milks with retinol. The average intake exceeded the RI for all age and sex groups (table 1). The proportion of the adult population (15–74 years) with intakes below 60% of the RI was 10% or less. The intake of β-carotene was on average 2 mg/day, slightly higher among women than among men. On an energy basis the women's diet contained more β-carotene than men's did. Smoking women had a lower daily intake of both vitamin A and β-carotene than nonsmokers, but this was correlated with a lower reported energy intake.

Vitamin E

The data on tocopherols are limited to intake of α-tocopherol. The vitamin E activity was calculated multiplying the intake of α-tocopherol with 1.2 according to the US RDA [3]. The average intake of vitamin E calculated in this way was above the RI for children and slightly below the RI for adults

Table 1. Daily intakes of certain vitamins among males and females aged 1–74 years: means and confidence limits (1.96 · SEM)

Age	n	Vitamin A mg ret.eq.		β-Carotene mg		Vitamin E[1] mg		Vitamin C mg		Folacin μg	
1–3 years											
Males	44	1.20	0.28	1.91	0.32	7.6	1.2	69	10	165	17
Females	34	1.14	0.28	1.21	0.28	5.9	0.8	58	11	152	19
NNR[2]		0.4		–		5		40		40	
4–6 years											
Males	56	1.21	0.16	1.73	0.31	6.8	0.5	71	9	177	13
Females	49	1.30	0.23	1.57	0.25	7.3	0.7	76	8	176	12
NNR		0.5		–		6		45		50	
7–10 years											
Males	53	1.46	0.21	2.20	0.56	7.9	0.7	85	13	218	20
Females	52	1.18	0.16	1.84	0.33	6.7	0.6	73	10	180	13
NNR		0.7		–		7		45		80	
11–14 years											
Males	55	1.30	0.20	1.84	0.28	9.1	0.9	89	16	247	24
Females	42	1.14	0.14	1.47	0.23	7.6	0.7	77	11	195	14
NNR		m 1.0	f 0.8	–		8		50		150	
15–18 years											
Males	56	1.55	0.18	2.02	0.44	9.7	1.3	90	16	265	21
Females	70	1.06	0.12	1.56	0.25	7.3	0.7	78	12	192	15
NNR		m 1.0	f 0.8	–		m 10	f 8	60		200	
19–24 years											
Males	79	1.38	0.15	1.55	0.22	9.6	0.8	76	12	243	15
Females	102	1.10	0.14	1.62	0.24	7.7	0.6	80	11	202	15
NNR		m 1.0	f 0.8	–		m 10	f 8	60		200	
25–64 years											
Males	565	1.60	0.09	1.81	0.11	9.2	0.4	69	3	224	6
Females	579	1.32	0.06	2.10	0.12	7.6	0.2	74	4	193	5
NNR		m 1.0	f 0.8	–		m 10	f 8	60		200	
65–74 years											
Males	109	1.56	0.17	1.98	0.25	8.9	0.8	70	8	214	12
Females	91	1.40	0.23	1.86	0.23	7.1	0.6	73	8	195	14
NNR		m 1.0	f 0.8	–		m 10	f 8	60		200	

[1] α-Tocopherol intake multiplied by 1.2.
[2] Nordic Nutrition Recommendations 1989 [2]. m = Males; f = females.

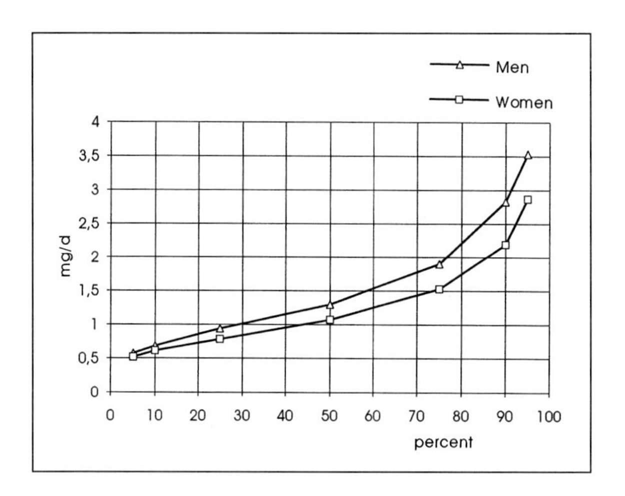

Fig. 1. Distribution of vitamin A intake (mg/day) among males and females aged 15–74 years in the Swedish HULK survey 1989.

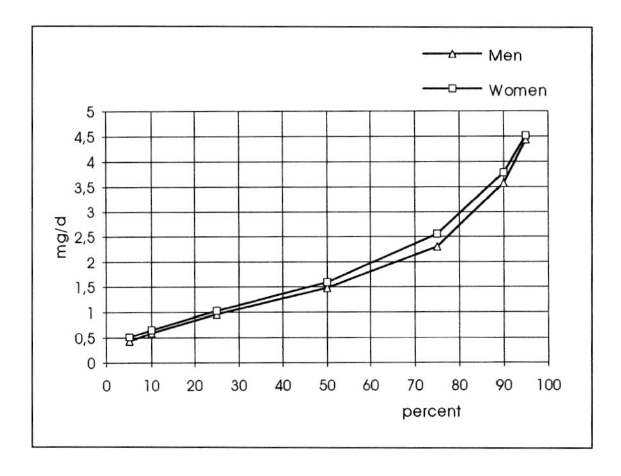

Fig. 2. Distribution of β-carotene intake (mg/day) among males and females aged 15–74 years in the Swedish HULK survey 1989.

(table 1). Approximately 15% of males and 10% of females 15–74 years old had an intake below 60% of RI (fig. 3).

Ascorbic Acid

The average intake of ascorbic acid was on average above the RI for all age groups (table 1). The data include corrections for estimated losses during

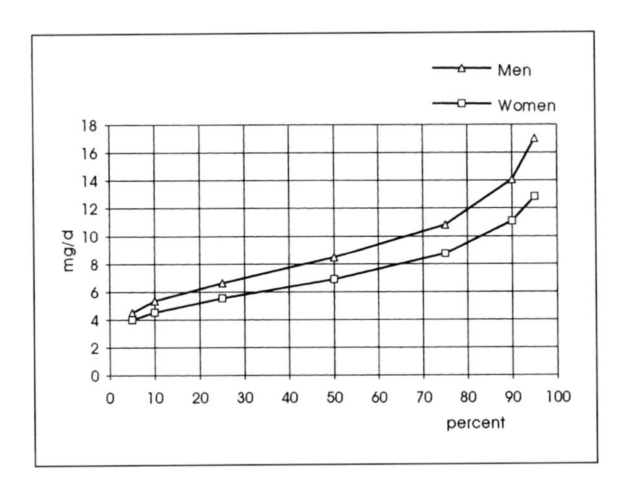

Fig. 3. Distribution of vitamin E intake (mg/day) among males and females aged 15–74 years in the Swedish HULK survey 1989.

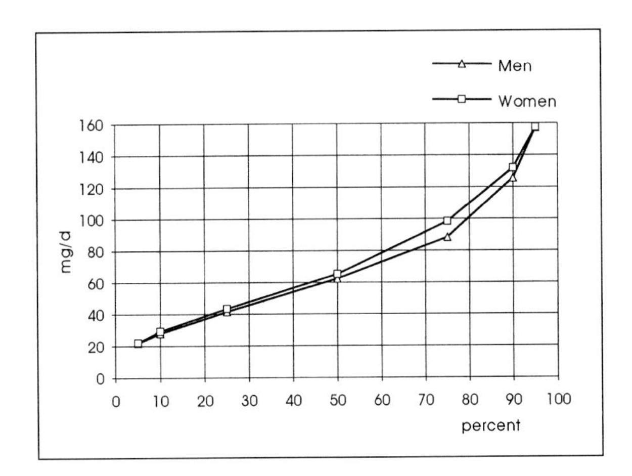

Fig. 4. Distribution of ascorbic acid intake (mg/day) among males and females aged 15–74 years in the Swedish HULK survey 1989.

preparation. However, about 10% of males had an intake below 60% of RI (fig. 4). On an energy basis the women's diet contained more ascorbic acid than the men's diet did. Smokers had a lower daily intake of vitamin C than nonsmokers. For men this difference remained when the vitamin C intake was adjusted for energy intake.

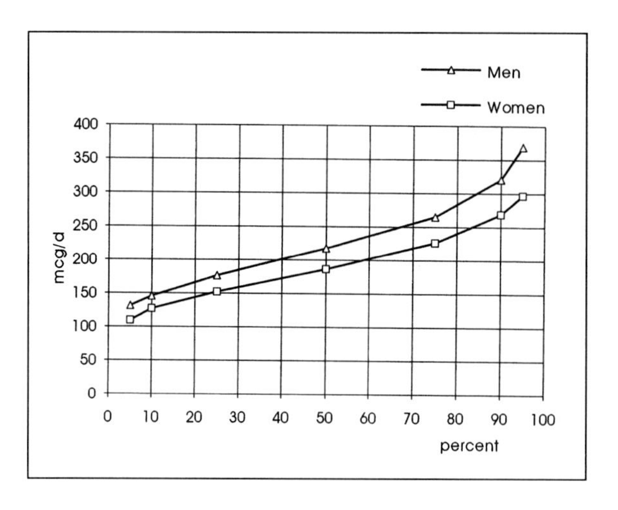

Fig. 5. Distribution of folacin intake (μg/day) among males and females aged 15–74 years in the Swedish HULK survey 1989.

Table 2. Intake and serum level of certain antioxidant vitamins among 50- to 69-year-old subjects in the Malmö Food Study 1984[1]

	Intake, mg/MJ		Serum level, μmol/l	
	men	women	men	women
Ascorbic acid	6.2	10.4	35.9	51.9
α-Tocopherol	0.89	1.07	33.2	36.1
Carotene	0.20	0.36	1.6	2.1

[1] From F. Lindgärde, B. Åkesson et al., unpubl. data.

Folacin

The average intake of folacin was above the RI for children and men, while it was close to the RI for women (table 1). Less than 5% of males and 5–10% of females 15–74 years of age had a intake below 60% of RI (fig. 5). In these figures, losses of folacin during cooking are not accounted for but generally the data for folacin in foods are less reliable. Smokers had a lower intake of folacin than nonsmokers, but this was correlated to a lower reported energy intake.

Biochemical Studies

There are few Swedish dietary surveys in which biochemical measurements of vitamin status have been included. In connection with the Malmö Food Study, serum levels of some antioxidant vitamins were measured in 206 50- to 69-year-old subjects who participated in a dietary survey in which food intake was measured using 18-day weighed food records [F. Lindgärde, B. Åkesson et al., unpubl. data]. In table 2 some preliminary results are presented. The correlation between the intake and serum level was for ascorbic acid 0.42 (women) and 0.45 (men) and for carotene 0.39 (women) and 0.26 (men). The data show that the differences in intakes between men and women are reflected in the serum levels.

Comparison with National Surveys in Other Nordic Countries

Nationally representative dietary surveys were carried out in Denmark in 1985 [4] and in Iceland in 1990 [6]. In Finland large dietary surveys covering certain regions have been carried out since the 1960s, the most recent in 1992 [7]. The Danish survey included 2,242 males and females 15–80 years of age, which constituted a representative sample of the adult Danish population. The response rate was ca. 80%. Information about the diet was obtained by a personal interview using the 'dietary history' method [4]. A similar method was used in the Icelandic study which included data on 1,240 males and females 15–80 years of age. The response rate was 72% [6]. The Finnish study in 1992 included data on 1,861 men and women 25–64 years of age from three geographical regions (East Finland, Southwest Finland and Helsinki area) covering the majority of the adult population in Finland. Information about the diet was obtained using a 3-day food record [7].

A direct comparison of the results from these studies has limitations due to use of different dietary survey methods, food composition tables, etc. However, a comparison of the nutrient density of the diet can give some information about the quality. In table 3 the dietary content of certain vitamins expressed per MJ among adults from these studies are shown. The data in table 3 indicate that the dietary content of vitamin A is somewhat higher in Iceland than in the other countries, while the dietary content of β-carotene is highest in Denmark and lowest in Iceland.

The content of vitamin E tends to be relatively similar, while the ascorbic acid content varies more. The higher level of ascorbic acid found in the Finnish survey can partly be explained by that losses due to preparation were not accounted for. In the Danish survey the estimated losses of ascorbic acid was

Table 3. Dietary content of certain vitamins (expressed per MJ) among adults in recent national Nordic dietary surveys

	Sweden[1] 15–74 years	Denmark[2] 15–80 years	Iceland[3] 15–80 years	Finland[4] 25–64 years
Vitamin A, mg				
Males	0.16	0.14	0.28	0.11
Females	0.17	0.17	0.29	0.11
β-Carotene, mg				
Males	0.19	0.26	0.07	0.21
Females	0.25	0.41	0.08	0.31
Vitamin E, mg				
Males	0.94	0.75	1.04	0.96
Females	0.97	0.96	1.19	1.02
Ascorbic acid, mg				
Males	8	5	6.5	13[5]
Females	10	7.5	10	19[5]
Folacin, μg				
Males	23	29	24	–
Females	25	33	28	–

[1] National survey 1989 [1].
[2] National survey 1985 [4, 5].
[3] National survey 1990 [6].
[4] National survey 1992 [7].
[5] Losses during preparation not accounted for.

on average ca. 30% on a daily basis. Applying this figure to the Finnish data gives an ascorbic acid content that is more close to what was found in the other surveys. The levels of folacin were relatively similar, with a tendency towards higher levels in the Danish diet.

References

1 Becker W: Food habits and nutrient intake in Sweden 1989 (in Swedish with English summary). Vår Föda 1992;44:349–362.

2 Nordic Nutrition Recommendations 1989, ed 2. Copenhagen, Nordic Council of Ministers, Standing Nordic Committee on Food, Report 1989;2.

3 National Research Council: Recommended Dietary Allowances, ed 10. Washington, National Academy Press, 1989.

4 Haraldsdottir J, Holm L, Højmark Jensen J, Møller A: Dietary habits in Denmark 1985. 1. Main results (in Danish with English summary). Publ No 136. København, Levnedsmiddelstyrelsen, 1986.
5 Pedersen JC: Folacin – New analytical results cause a re-evaluation of the folacin intake in Denmark (in Danish). Scand J Nutr 1988;32:113–116.
6 Steingrímsdóttir L, Þorgeirsdóttier H, Ægisdóttir S: Dietary habits in Iceland 1990. 1. Main results (in Icelandic). Reykjavik, Islands Ernæringsråd, 1991.
7 Kleemola P, Virtanen M, Pietinen P: Dietary survey of Finnish adults in 1992. Publications of the National Public Health Institute, B2/1994.

Dr. Wulf Becker, Nutrition Division, National Food Administration,
PO Box 622, S–751 26 Uppsala (Sweden)

Walter P (ed): The Scientific Basis for Vitamin Intake in Human Nutrition.
Bibl Nutr Dieta. Basel, Karger, 1995, No 52, pp 146–157

Criteria and Scientific Basis for RDA (PRI)

W. Philip T. James

Rowett Research Institute, Aberdeen, UK

Since this symposium deals with vitamin needs, only these will be considered except insofar as the designation of a vitamin requirement is affected by the energy and protein requirement figures or by the intake of a macronutrient. This is particularly important this year as the Scientific Committee for Food (SCF) in the European Union has just issued a new set of reference values taking account of many different national attitudes to the process by which these are calculated [1].

Recent Developments in European Nutrition Policy-Making

For those who are unfamiliar with the European bureaucracy, the SCF is an independent advisory committee to the Commission itself and is not a body of nationally elected representatives whose task it is to reflect different Government views. In practice the SCF was restructured in 1992 with the addition of new members. Those chosen were selected from either a short or a long list of nutritional experts submitted by national Governments. The Commission then chose their experts bearing in mind both the need to cover a wider spread of issues than those dealt with hitherto, e.g. microbiological issues, molecular biology and more extensive nutritional expertise but at the same time preserving where possible a national balance so that politically the Commission could not be charged with bias. The SCF was extended in 1992 to 15 members and recently this membership was expanded further by designating a series of subgroups to which further specialists are being invited. This expansion of activity was instituted following the Commission's concern for standardizing food legislation throughout the European Union. This is needed if free trade and comparable food standards are to be developed.

The whole of food legislation became a high profile political issue during the Edinburgh Summit meeting of National Government leaders where the host nation, the United Kingdom, sought to display the Commission's obsession with unnecessary legislative detail by highlighting the plethora of regulations affecting food. These political developments led in the Maastricht Treaty to the doctrine of subsidiarity where it was considered appropriate to transfer back to national authorities tasks which were best conducted at the national level. Part of this process has involved food legislation where the idea is to allow a country with particular expertise, e.g. in nutritional surveillance to review, summarize and propose approaches which the Commission could then enact. These national groups of experts in Institutes, working individually or as part of a governmental process, were considered best able to reduce the load on the SCF which was in danger of being overwhelmed by work. A national process usually involves scientifically trained civil servants, often with remarkable skills and expertise. They integrate expertise, undertake preliminary screening and liase with industry, consumers, legal, trade and financial experts within Governments while providing a secretariat for the committees. In the Commission, however, there is a remarkably small pool of civil servants – contrary to popular belief, so small as to be comparable in size to that in the Scottish Office. Thus there is a rational for devolving the SCF's work to national Governments but the SCF is anxious about the inevitable perceived need to share the burden with all constituent members of the European Community. Members fear that they will then find themselves embarrassed when, for political reasons, a government appoints an incompetent expert to advise the whole European Union on an issue of some importance. There is thus a jostling for position between the politicians and national governments, seeking to limit the power of the Commission on every issue and the Commission's jealous guarding of its capacity to propose new approaches using such independent advice as symbolized by the SCF as it can get.

This summary of political developments may seem surprising in this analysis of European principles underlying vitamin requirements but it is highly relevant for those who believe that recent reports are either biased for political reasons or need extension or change. The SCF is indeed independent but in the future the process by which one might amend or extend current vitamin recommendations is very unclear.

The SCF Approach to Vitamin Requirements

The mandate given to the SCF by the Commission was to advise on the establishment of European Recommended Dietary Allowances for a number of purposes including nutrition labelling and Commission programmes on research and nutrition and to make recommendations. The SCF responded by recognizing that almost all countries of the European Community (as it then was) had independently convened a committee of experts to derive value for nutritional recommendations. Although reasonably similar in their conclusions the various reports differed in emphasis because different groups of experts are inevitably going to come to somewhat different conclusions, particularly when the data available is often sparse.

Rather than simply conducting a paper exercise to establish where the differences lay, the SCF established a working group of 19 experts to re-examine the evidence. The experts were divided into four working parties to cover (a) energy and protein, (b) water-soluble vitamins, (c) fat-soluble vitamins and (d) minerals and trace elements. Their conclusions involved a renewed scrutiny of the most up-to-date information and a recognition of the various strategies adopted by nutritional and other committees external to the European Union.

The Derivation and Use of RDAs

Traditionally an RDA has been generated from physiological, biochemical and epidemiological data to reflect not the average need of a group or population but that level which encompasses all or nearly all the population's needs. Where this is somewhat speculative – a frequent occurrence – the value is estimated from physiological and biochemical studies which provide an average value for the appropriate intake of a nutrient and an indication of the variability in the values between individuals. Since this variability can be approximated crudely to a Gaussian distribution then a value corresponding to the mean +2 SD of the mean will, in theory, cover 97.5% of the population. In the many cases where the validity of the standard deviation value is unknown a 15% SD figure is chosen and frequently a further safety margin is added. This classic approach and the choice of the phrase *Recommended Dietary Allowance* was viewed with some scepticism by the SCF because not only did the safety margins vary but the name led to considerable misinterpretations. The conclusion was reached that the average requirement, often the most clearly defined from physiological studies, should be separately designated. The concept of a variability within a group should then be reflected in the upper level of 2 SD by designating it the Population Reference Intake (PRI). Finally, the mean –2 SD figure was also listed as the Lowest Threshold Intake (LTI) to indicate that below this value almost everybody's nutrient intake could be expected to be inadequate.

It could be argued that these changes are cosmetic or simply reflect the SCF's need to appear independent and different. The emphasis on the choice of three values and the changes of name seemed to the author to reflect the exasperation of the working groups that for many commentators, scientists, doctors and even nutritionists were misinterpreting intake data and policy decisions by assuming that any individual whose intake was below the RDA was intrinsically deficient. The probability theory, so common to public health considerations, is evidently still mysterious, not only to consumers but to academics most of whom are trained to provide and interpret precise and assured data in laboratory experiments.

The Basis for Calculating Average Nutrient Requirements

Presumably every reader is familiar with the fact that requirements are based on physiological and biochemical criteria which are deemed to relate to a function perceived as important in public health terms. It is difficult to make a case for the SCF being particularly innovative in these processes particularly as far as the vitamins are concerned. In terms of iron needs, however, Hallberg exerted a remarkable influence consonant with his lifetime involvement in the field. The SCF's unusual propositions are covered elsewhere in this symposium by Hallberg himself. Perhaps the principal highlight relates to the European need to consider vitamin D requirements in view of the concerns for bone disease in the elderly and the limited amounts of vitamin D synthesized in the skin of populations living in northern latitudes. The other small difference relates to the dependence of some vitamins on the expected energy requirement where again some new European approaches to the calculation of energy needs were adopted. The SCF certainly took the classical, cautious approach to the process of assessing vitamin needs so the report showed the expected dichotomy of approach when dealing with the fat-soluble and water-soluble vitamins. This in part reflects the huge difference between the two vitamin groups with limited storage of water-soluble vitamins and therefore the enhanced susceptibility to their early depletion. One of the unusual features of the report was the emphasis on highlighting discrepancies and concerns rather than implying that the experts were happy with the data. This will be illustrated by considering vitamin A in some detail since some of my analyses of the problems are very relevant to this meeting.

Vitamin A and the Carotenoids

The linking of clinical features of vitamin A deficiency in American volunteers with estimates of body pool size, liver concentrations and plasma levels of retinol, was complicated by the fact that plasma retinol levels in North Americans and Europeans are much higher at a given body retinol pool size than the plasma levels found in Thais. Thus plasma levels which seem appropriate for a European may not be suitable in Thailand. Plasma retinol concentrations are maintained at a reasonably constant level in an individual and in the UK average plasma levels are 63 μg/dl (2.2 μmol/l) in men and 54 μg/dl (1.9 μmol/l) in women. However, in apparently healthy Thais their plasma levels are maintained below 30 μg/dl (1 μmol/l) even when liver stores are quite high [2]. This presents problems in conceptual terms because if, as Blomhoff et al. [3] have shown so elegantly, there is a homeostatic mechanism which delivers retinol from the special stellate storage cells in the liver to the hepatically derived retinol binding protein which then serves as the carrier of retinol in the blood, one has to ask why there should be such big differences between UK and Thai values. Since the circulating retinol is the means by which the peripheral tissues are thought to obtain retinol, presumably the plasma concentration or perhaps the plasma flux of retinol is the key process depicting adequacy. Yet all current methods of calculating retinol requirements have been based on obtaining with labelled retinol a fractional catabolic rate and a measure of body pool size while human volunteers are being depleted or repleted. Olson [5] had proposed pragmatically that vitamin A sufficiency required a liver concentration of 20 mg retinol or its equivalent in the esterified form per gram wet weight of liver since no clinical sign of deficiency had been noted in subjects with levels greater than this and this liver concentration could maintain a steady-state concentration of retinol at plasma levels in excess of 20 μg/dl (0.7 μmol/l). Furthermore, this liver concentration was known to be adequate for maintaining an adult on a vitamin-free diet for months without exhibiting deficiency.

What is not known is why the Thais have such a low retinol concentration in their blood at equivalent levels of hepatic stores of retinol. Perhaps one explanation is the well-recognized lower protein intake in Thailand. The retinol binding protein concentrations might be expected to be much lower because we showed many years ago the extraordinary susceptibility of this protein to both energy and protein depletion [4]. This, of course, does not mean that the Thais are necessarily at a disadvantage with these low values because the supply of retinol to the tissues will then depend on the turnover rate of retinol pool in plasma. A further explanation may, however, prove to be of greater importance and relevance to this Symposium. One can readily assume, as

Olson [5] has noted, that in Thailand the demand for retinol is met for the most part by the conversion of carotenoids to retinol. This was principally thought to occur by central cleavage of β-carotene and other carotenoids in the intestine. Now Russell and co-workers [6] have demonstrated an alternative pathway whereby peripherally distributed carotenoids can generate retinol by a newly discovered eccentric cleavage enzyme. We therefore need to take account of a potentially novel approach to carotenoids. Therefore, since usual central cleavage of carotenoids in the intestinal mucosa leads to retinol transport to the liver, the eccentric cleavage of carotenoids in peripheral tissues seems to provide a completely different supply of locally derived retinol. It seems quite possible therefore that in the Third World it is not so much the retinol linked to RBP circulating in plasma which is important for supplying retinol to the periphery, but the peripheral concentrations of the carotenoids which provide a separate route for the local formation of vitamin A. If this concept is substantiated then Blomhoff's discovery of the tightly controlled metabolic regulation of retinol transport from the liver into the plasma may be seen as an evolutionary appropriation system for limiting vitamin A toxicity, whilst being capable of providing the tissues with a reserve supply of retinol. Given the dependence of Third World communities on the carotenoids for providing retinol equivalents, we may need to rethink the issue of the linkage between vitamin A and carotenoid requirements in Europe and not be dominated in our thinking by data from American or British subjects with their high vitamin A consumption. This issue may also have substantial relevance for Mediterranean societies with their much higher carotenoid intakes.

The SCF did consider the carotenoids and β-carotene in particular, in relation to both retinol supply and as an antioxidant. It was recognized that intakes or plasma levels of β-carotene were associated with a lower rate of a variety of cancers but of lung cancer in particular [7]. The problem is whether β-carotene is simply a marker for other carotenoids or important in its own right. The very recent [8] findings of an enhanced risk of lung cancer in smokers taking extra β-carotene highlight the danger of assuming β-carotene is the principal effector of carotenoid action. The use of a pill with β-carotene given as a single dose may be disadvantageous because it could well inhibit the uptake of other carotenoids. Therefore, by being so readily converted to retinol in the intestine, the potential antitumour, antioxidant effects of β-carotene may be limited and paradoxically a single bolus of β-carotene may be disadvantageous if the other carotenoids supplying peripheral tissue antioxidants are impaired in their absorption.

Vitamin E

This issue was again considered in somewhat cautious terms by the SCF. The classic deficiency syndrome of neurological, retinol and skeletal changes which appears after years of eating a deficient diet does not seem relevant to the other important functions of vitamin E as an antioxidant. Nor does the specific syndrome of haemolytic anaemia, thrombocytosis and oedema in deficient premature babies seem relevant. Therefore, new approaches are being developed based on vitamin E's antioxidant effects. A red cell test which assesses erythrocyte resistance to an artificial oxidative agent reveals that membrane resistance is substantial if blood α-tocopherol (α-TC) levels are >2.25 μmol α-TC/μmol cholesterol. As PUFA intakes rise so does the dietary intake of α-TC needed to maintain the α-TC/cholesterol ratio at 2.25 μmol/μmol. The working party of the SCF considered whether to take the American assessment of the appropriate dietary ratios of α-TC and PUFAs as 0.4 mg α-TC/g PUFA [9, 10]. On this basis a recent UK study showed that 99% of the British population have a α-TC/cholesterol level of over 2.25 and this implied that the 95th percentile adult intake of α-TC should be 19.5 mg α-TC equivalents/day in men and 15.2 in women. This assessment, however, had to take account of the PUFA intake which varies markedly in Europe. Since 0.4 mg α-TC/g PUFA was the observed ratio in a normal American diet, and Americans have a lower rate of heart disease than that found in the UK, this led to an acceptance of the American dietary ratio. The higher American value semed justified because even on low PUFA intakes there would still be a need for α-TC. An LTI of 4 mg α-TC/day in men and 3 mg/day in women was considered adequate.

In our recent studies of the European diet we have observed a wide range of α-TC 'intakes' (7–24 mg α-TC supply/capita/day) [10]. These intakes are admittedly based on food balance sheets, but these were corrected for a series of errors once we had examined in detail the potential contributors to α-TC dietary supplies. Thus in those Mediterranean countries with very low rates of coronary heart disease, the average per caput amount of α-TC supply in the diet was 19 mg [11].

Managing Issues Relating to the Antioxidants

This workshop might reasonably ask why the SCF took such a conservative view of the requirements for antioxidants. The problem in essence is how best to develop a whole-body approach to the assessment of antioxidant turnover whilst taking account of the differential needs of particular organs. This can be illustrated by a series of examples.

First, it is still not clear how the fat-soluble and water-soluble system of antioxidants are integrated in quantitative turnover terms. We do not know, for example, whether in the fat-soluble system, ubiquinone can substitute for the carotenoids in the chain-breaking process of free radical scavenging within the lipid phase. Since β-carotene is more effective at the low partial pressures of oxygen found in mammalian tissues [12] than α-TC, what should the relationship be between carotenoid or ubiquinone needs and α-TC requirements? Recent data from a European study highlighted the greater predicted benefit of β-carotene than α-TC levels measured in adipose tissue, in supposedly protecting against coronary heart disease [13] whereas we found a much more powerful protective effect of dietary supplies of α-TC [11]. The natural tendency would be to give greater credence to the β-carotene link based on direct measures in the adipose tissue of randomly sampled adult males. These data seem more reliable than information based on food supply data collated on a national basis. However, some caution is in order because we are still unclear whether the adipose tissue concentrations of β-carotene and α-TC do reflect on an equivalent basis the dietary intake or even whether they predict the same β-carotene/α-TC ratios in tissues other than adipose tissues. Dutta-Roy et al. [14] at the Rowett Research Institute in Aberdeen, have discovered a novel 14.5-kD α-TC-specific transport protein in heart and other tissues which can now explain the unusual inter-organ distribution patterns of the different isomers of vitamin E. This protein is avid for α-TC but not for γ-TC. Dutta-Roy et al. also found an uptake protein for α-TC in liver membranes. The 14.5-kD protein for α-TC transport is also present in the liver as well as the 30-kD protein which was described earlier by Catignani and Bieri [15]. The implications of these findings are dealt with elsewhere [16] but in essence we need to be cautious about interpreting the β-carotene/α-TC relationships in adipose tissue as indicative of tissue levels if we now find selective and specific organ uptake and transport of selective isomers of vitamin E. Thus α-TC levels may be much higher in cardiac tissue than that expected from their levels in adipose tissue. The fact that γ-TC is rapidly cleared from the blood to the liver has been taken as indicative of the minor importance of γ-TC compared with α-TC. Yet the biological activity of γ-TC exceeds that of α-TC. I would argue that the one tissue with extraordinary oxidative activity is the liver with its major detoxification processes, very high metabolic rate and huge influx of portal bacteria which need removal, killing and processing. Thus in biological terms it is the liver which needs the maximum level of free radical scavenging. On this basis one could consider the rapid clearance of γ-TC from the blood as a mechanism for selectively channelling the most effective antioxidant, γ-TC to the principal site of oxidant activity in the body. Thus I would claim that those who concentrate on the significance of α-TC for the prevention of cardiovascular disease may well be right but the SCF will also need to consider whether we have been neglecting the biological role of γ-TC in preserving the integrity of liver function. This may be especially important in those on a high alcohol intake when free radical overload may contribute to the pathogenesis of alcoholic liver diseases [17].

Water-Soluble Antioxidants

The role of glutathione as a free radical scavenger has been highlighted by Golden and Ramdath [18] who consider the pathophysiological effects of kwashiorkor to be one of free radical-induced liver damage associated with high liver iron stores (which are perhaps enhanced by haem compounds of bacterial origin). Glutathione levels are low but much less is known about the ascorbic acid status. Vitamin C seems to occupy a central position in antioxidation by acting as an acceptor of free electrons transferred from both fat- and water-soluble antioxidants. Turnover is known to be higher in smokers than in nonsmokers [19], but it is unclear to what extent ascorbate turnover can be seen as the final common pathway of oxidative turnover and an index of the body's burden of free radical damage.

These scientific issues are important if the SCF is to reconsider how best to handle the challenge of specifying the PRIs of these nutrients. If to this problem we then add the increasing evidence that flavonoids, derived from tea and other foods, may be protective for cardiovascular disease as may the flavonoids and other bioactive molecules to be found in red wine [19, 20], then we are moving into a complex and new approach to antioxidant needs.

Folic Acid

The SCF was very troubled by the issue of folate requirements in view of the reasonable proposition that periconceptual intake of 400 mg folic acid, in addition to that found in the diet, protects against neural tube defects. The issues of concern were the question of whether these women were unduly susceptible to NTD because of a definable genetically determined change in folate metabolism, e.g. by having an abnormal cystine synthetase. This might provide a strategy for screening eventually but meanwhile the concern was that if a PRI or RDA were proposed at a much higher level for folate then an increasing proportion of the elderly might have their vitamin B_{12} deficiency exacerbated with the development of irreversible spinal atrophy of the dorsal and lateral columns. For this reason the SCF was again conservative and considered the 400 μg of folate as a special need for women of childbearing years.

Other Vitamin-Related Issues

Two other unusual developments in the SCF's considerations have a tangential bearing on some vitamin measures. These are those that relate to body weight and to the energy needs in pregnancy.

For body weight-related data, new information was obtained on the average body weight of adult men and women in Europe. This was generated from a long series of national data sets so that with appropriate weighting for the different population numbers reasonably valid weights, heights and BMIs were obtained. These values were then used to obtain energy requirement estimates. These are used for calculating the need for some water-soluble vitamins.

The second minor feature was a new specification, based on US data, of different energy needs of pregnant women who start pregnancy at different BMIs and have different desirable weight gains during pregnancy. The desirable weight gains were based on the need to produce babies of 3.5 kg or more. These different weight gains were then used to calculate different extra energy needs with or without an allowance for changes in physical activity.

Choice of Value for Labelling Purposes

This was given considerable thought by the working party which finally determined that the Average Requirement (AR) for men should be chosen for complex reasons. First they were concerned that an RDA type value was being misused to imply the need for micronutrient supplementation. Secondly many adults were beginning to consider themselves as deficient when they tried to relate these RDAs to their personal intake. Thirdly, even normal and nutritionally valuable foods would tend to be excluded from specifying their nutrient content since it is required that 15% of an RDA had to come from 100 g of a particular food before it can be considered to contain sufficient amounts of the particular nutrient to warrant its inclusion on a label. By designating the reference value for labelling as the average requirement for men and identifying it as an average, then women, who do the shopping for the most part, would automatically be covering their own needs and those of their children and elderly relations by relating to this value. This figure would also not increase concern that a diet was of poor quality.

This compromise was also chosen because a single value was needed for labelling purposes. Only one micronutrient was excluded from this rule and that was to take account of the unusually high iron needs of menstruating women. In this case the labelling value was proposed corresponding to the level needed to cover the iron needs of 80% of premenopausal women. This extra value would be placed in brackets after the AR value for men.

Conclusion

Nutritional science is blossoming once more and the attempts by national or European Union communities to specify an appropriate diet for long-term health is becoming ever more difficult. The issue of antioxidants is also becoming very complex and it seems likely that over the next decade an integration of nutritional physiology, cell biology and epidemiology will allow us to produce a more coherent overview of the interplay between dietary bioactive molecules and the relative amount of antioxidant micronutrients. Once this understanding emerges then I foresee the need to revise – perhaps substantially – this first attempt by the SCF to produce European figures for the requirements for nutrients.

References

1 European Communities – Commission: Nutrient and energy intakes for the European Community. Reports of the Scientific Committee for Food (31st Series) 1993.
2 Suthutvorvost S, Olsen J-A: Plasma and liver concentrations of vitamin A in a normal population of urban Thais. Am J Clin Nutr 1974;27:883–891.
3 Blomhoff R, Green MH, Berg T, Norum KR: Transport and storage of vitamin A. Science 1990; 250:399–404.
4 Shetty PS, Watrasiewicz KE, Jung RT, James WPT: Rapid-turnover transport proteins: An index of subclinical protein-energy malnutrition. Lancet 1979;ii:229–232.
5 Olson JA: Needs and sources of carotenoids and vitamin A. Nutr Rev 1994;52:S67–S73.
6 Wang XD, Krinsky NI, Tang GW, Russell RM: Retinoic acid can be produced from eccentric cleavage of beta-carotene in the human intestinal mucosa. Arch Biochem Biophys 1992;293:298–304.
7 Zeigler RG: Vegetables, fruit and carotenoids and the risk of cancer. Am J Clin Nutr 1991;53: 2515–2595.
8 The Alpha-Tocopherol, Beta-Carotene Cancer Prevention Study Group: The effect of vitamin E and beta-carotene on the incidence of lung cancer and other cancers in male smokers. N Engl J Med 1994;330:1029–1035.
9 Bieri JG, Everts RP: Tocopherols and fatty acids in American diets: The recommended allowance for vitamin E. J Am Diet Assoc 1973;62:147–151.
10 Witting LA, Lee L: Dietary levels of vitamin E and polyunsaturated fatty acids and plasma vitamin E. Am J Clin Nutr 1975;28:571–576.
11 Bellizi MC, Franklin MF, Duthie GG, James WPT: Vitamin E and coronary heart disease: The European paradox. Eur J Clin Nutr 1994;48:822–831.

12 Burton GW: Antioxidant action of carotenoids. J Nutr 1989;119:109–111.
13 Kardmaal AFM, Kok FJ, Ringsted J: Antioxidants in adipose tissue and risk of myocardial infarction: The EURAMIC Study. Lancet 1993;342:1979–1984.
14 Dutta-Roy AK, Gordon MJ, Livingston DJ, Paterson BJ, Duthie GG, James WPT: Purification and partial characterisation of an α-tocopherol-binding protein from rabbit heart cytosol. Mol Cell Biochem 1993;123:139–144.
15 Catignani GL, Bieri JG: Rat liver α-tocopherol binding protein. Biochim Biophys Acta 1977;497: 349–357.
16 Dutta-Roy A, Gordon M, Campbell FM, Duthie GG, James WPT: Vitamin E requirement, transport and metabolism: Role of α-tocopherol binding protein. J Nutr Biochem 1994;5:562–570.
17 Leiber CS: Biochemical and molecular basis of alcohol-induced injury to liver and other tissues. N Engl J Med 1988;319:1639–1650.
18 Golden MHN, Ramdath DD: Free radicals in the pathogenesis of kwashiorkor. Proc Nutr Soc 1987;46:53–68.
19 Duthie GG, Arthur JR, Beattie JAG, Brown KM, Morrice PC, Robertson JD, Shortt CT, Walker KA, James WPT: Cigarette smoking, antioxidants, lipid oxidation and coronary heart disease. Tobacco smoking and nutrition. Ann N Y Acad Sci 1993;686:120–129.
20 St Leger AS, Cochrane AL, Moore F: Factors associated with cardiac mortality in developed countries with particular reference to the consumption of wine. Lancet 1979;i:1017–1020.

W. Philip T. James, Director, Rowett Research Institute, Greenburn Road,
Aberdeen AB2 9SB (UK)

Walter P (ed): The Scientific Basis for Vitamin Intake in Human Nutrition.
Bibl Nutr Dieta. Basel, Karger, 1995, No 52, pp 158–167

Discussions in the United States about Recommended Daily Dietary Allowances in the Future

Paul A. Lachance

Department of Food Science, Rutgers, State University of New Jersey,
New Brunswick, N.J., USA

At the height of World War II, the National Academy of Science, National Research Council (NAS/NRC) published a 6-page document entitled 'Recommended Dietary Allowances' [NAS, 1943]. Since then, the RDAs, in 10 successive editions, have become the most authoritative guide to recommended nutrient intakes in the United States, and have been used for a variety of purposes. In anticipation of the 11th edition of the RDAs, the Food and Nutrition Board of the NAS/NRC solicited input to the following queries:

What has been the experience in applying the RDAs in various settings, and what factors limit their use?

What new evidence has arisen since publication of the 10th edition of the RDAs that would argue for a change from the present values or a re-examination of the evidence?

Should concepts of chronic disease prevention be included in the development of allowances? For which nutrients and other food components?

How should recommended levels of intake be expressed? Should single numbers be given for different age and sex categories, or should ranges of recommended intake be provided? How should the ranges be defined? Should toxic levels be included where data are sufficient to establish an upper acceptable limit?

Is knowledge of relationships among nutrients sufficient to consider when establishing RDAs?

These questions were the basis for a workshop on the future of the RDAs held at Rutgers University on April 13, 1993 [Lachance and Langseth, 1994].

The principal focus however was on the crucial question of whether concepts of chronic disease prevention should be included in the development of RDAs.

US Consumers Receive Differing Types of Advice

The American public receives two distinct types of nutrition advice. The NRC report 'Diet and Health' [NAS/NRC, 1989a], the report 'Dietary Guidelines for Americans' [USDA/DHHS, 1990] and the 'Surgeon General's Report on Nutrition and Health' [DHHS, 1988] all emphasize macronutrients, especially dietary fats, and overall dietary patterns, and their impact on the risk of chronic diseases. The second type of advice consists of food labeling standards derived both from the RDAs and the aforementioned dietary guidelines. Micronutrients and the prevention of classic deficiency syndromes were readily apparent in the 1973 Food Labeling regulations, whereas with the advent of the Nutrition Education and Labeling Act (NLEA) in 1992, the emphasis has shifted to macronutrients at the expense of micronutrients. Neither set of recommendations addresses a topic that has been a key focus of nutrition research for more than a decade – the relationship between micronutrient intake and the prevention of nondeficiency diseases. All but ignored are the new and valuable insights in nutrition science, such as the link between folic acid and neural tube defects, as well as colorectal cancer and elevated serum homocysteine in coronary heart disease; the role of calcium and other micronutrients in the prevention of osteoporosis, hypertension and possibly colon cancer; and the potential protective effects of antioxidant micronutrients against a number of cancers, cardiovascular disease and degenerative diseases of the eye.

Consider, for example, the impressive evidence linking β-carotene and other carotenoids with the reduced risks of several cancers [Gaby and Singh, 1991]. In the 1989 RDA document, the committee acknowledged that 'a generous intake of carotenoid-rich foods may be of benefit' to health [NAS/NRC, 1989b]. But the lay consumers are exposed to the 'disease of the week' by the electronic and print media and to the disagreements amongst scientists rather than the direction of the consensus that is occurring. For example, whereas there is no specific RDA for carotenes (one half of the vitamin A RDA is said to be derived from precursor plant food carotenoids), the text of the 'Dietary Guidelines for Americans' urges 'Choose a diet with plenty of vegetables, fruits and grain products' but does not mention the word 'carotenes'.

The Food Guide Pyramid has replaced the Basic Four Food Groups scheme of education, yet the pyramid does not specifically call for the inclusion of deep yellow/orange or leafy dark green vegetables in the daily diet, these foods are the major sources of carotenoids, tocopherols, ascorbic acid and folic acid. The format of the newly mandated (May 8, 1994) nutrition label does not allow a distinction to be made between carotenes and preformed vitamin A.

There appears to be no place in the current two-tiered scheme of dietary guidance for promoting the emerging knowledge that fosters and explains the micronutrient disease prevention characteristics of antioxidants and free radical scavengers. The dietary goals for chronic disease prevention do not include nutrient goals. The RDAs in their current representation only address prevention of deficiency and are not directed to identify more optimal intakes for chronic disease prevention. The 1989 RDA document states 'it is not possible at this time to establish optima'. The Rutgers workshop and the EANS workshop represent efforts to recognize the role of micronutrients in chronic disease prevention, and to suggest possible levels of intakes that thwart chronic disease and that differ from levels intended to prevent specific nutrient deficiencies.

A Shift in Philosophy

Weisburger [1991], amongst others, has proposed the concept that the RDAs for nutrients be designed not only for the avoidance of chronic diseases but also for a more optimal protection against environmental toxicants. Interestingly, the 1989 RDA and the Canadian RDI [Health and Welfare Canada, 1990] took a step in that direction by acknowledging that cigarette smokers have lower serum levels of ascorbic acid. Smoking (possibly the most concentrated source of individual environmental toxicants) appears to increase the metabolic turnover for energy [Hofstetter et al., 1986], as well as for ascorbic acid per se [Kallner et al., 1981]. The nutrient prevention approach that is recommended must however be comprehensive and not based on evaluation of single nutrients as if they were drugs. The human body is approximately 70% water and composed of 100 trillion cells. The first line of defense to oxidative insults in the water-soluble antioxidant ascorbic acid and the enzymatic processes (e.g. glutathione reductase, superoxide dismutase, catalase) that process oxidative constituents. Lipid-soluble antioxidants incorporated in the various lipid bilayer membranes of the cells are second and third lines of defense to oxidative insult [Esterbauer et al., 1992; Jialal and Grundy, 1993]. Human intervention studies of the long-term oxidative insults of smoking [Heinonen et al., 1994] which fail to incorporate and study the limiting antioxidant – ascorbic acid – serve to demonstrate the pre-existence of disease rather than the prevention of disease and falsely incriminate tertiary antioxidants such as β-carotene.

Dietary or Nutrient Allowances

The term Recommended *Dietary* Allowances has the implication that the desired amounts of nutrients can and should be obtained from a diet of nominal foods. As energy expenditure has decreased due to conveniences and more sedentary lifestyles, obesity has increased [Lachance, 1994a,b] and the recognition exists that reducing diets must either include nutrified foods or a multivitamin/mineral supplement. Food fortification contributes 4–20% of nutrients to the dietary [Lachance, 1989]. Certain 'optimal' intakes may be difficult or impossible to obtain from diet alone. Thus in the future, recommended *nutrient* allowances may need to be distinguished from *dietary* allowances.

Folic acid is a case in point. Most experts agree that women of childbearing potential should consume approximately 400 μg/day in order to reduce the risk of bearing a child with a neural tube defect [CDC, 1992, 1993; Rosenberg, 1992]. Older persons may need similar intakes to maintain normal plasma homocysteine levels which is associated with a decreased incidence of a myocardial infarction [Selhub et al., 1993; Stampfer et al., 1993]. Yet about a third of Americans fail to meet the 1989 RDA of 200 μg, because most do not consume the recommended number of servings of fruits and vegetables daily [Block, 1991]. An ideal diet would provide about 367 μg [Lachance, 1992]. Consumers would benefit from the restoration of cereal grain products with folic acid [NAS/NRC, 1974]. Some nutritionists have difficulty retreating from the established dogma that diet is sufficient and that fortification and supplement use are unnecessary. Yet, the fortification of grain products with thiamine, riboflavin and niacin has thwarted the re-emergence of the associated deficiency diseases for decades. Further, supplement users appear to have a lower overall and cardiovascular disease mortality [Enstrom, 1992], and the calculated impact of ascorbic acid, tocopherol and carotene antioxidant use is a 10% reduction in health care costs [Anonymous, 1993]. In the case of vitamin E and cardiovascular disease [Stampfer et al., 1993; Rimm et al., 1993] it may be difficult to obtain the 'optimal' amount that affords protection from diet alone. If further research confirms the epidemiological evidence and the clinical trial studies of the protective effects of E on lipoprotein oxidation [Reaven et al., 1993; Jialal and Grundy, 1992], it may be necessary to conclude that at least 100 mg/day are needed to afford protection against atherosclerosis. Such intakes cannot be readily and routinely obtained from even 'ideal' diets without concomitantly increased energy intakes.

Selecting Values for a Reconceptualized RDA

Selecting recommended intake values that would decrease the risk of chronic diseases creates a reluctance in some nutritionists. Even for nutrients where most agree more ample intakes are desirable, such as the antioxidant nutrients, selecting a single number is problematical. But, choosing a single number has always been problematical. The presence of specific nutrients values in an RDA table gives an illusion that is not truly justified. A close reading of the RDA document reveals that there is considerable uncertainty in the 1989 allowances, even though prevention of chronic diseases was not taken into consideration. For example, the 1989 RDA acknowledged that ascorbic acid allowances were set 'somewhat arbitrarily'. For calcium, one finds the statement that an optimal intake 'is difficult to define'. For zinc, the setting of an RDA was 'beset with several uncertainties'. The vitamin E RDA is based on the 'arbitrary but practical' value of customary intakes.

Since the values for preventing chronic diseases are invariably higher than the deficiency disease values, the task should not be daunting unless one is paranoid about toxicity, also an issue the RDAs have avoided discussing. The majority of nutrients in the RDA have no known role in chronic disease prevention. For these, no radical change in the method of establishing the RDA is apparently needed. For other nutrients, there is already agreement on levels of intake that would decrease the risk of chronic diseases and these approach the values calculated for 'ideal' diets (table 1) [Lachance, 1992]. Diets of this type have been associated with the reduced risks of cardiovascular disease and cancer [Kant et al., 1993], and the antioxidant nutrients present in these diets are believed to contribute risk-reducing effects. These values could readily be used as a starting point (fig. 1–3). The RDA for folic acid was 400 μg for more than 20 years, and was only decreased to 180–200 μg in the 1989 RDAs. There is also reasonable agreement on calcium at 1,000–1,500 mg/day [NIH, 1984]. An updated calcium consensus conference is scheduled for June 1994, where emphasis will be placed on the role of vitamin D and other nutrients in addition to calcium alone on bone metabolism and the prevention of osteoporosis. This is an important move away from the tunnel vision association of only one nutrient with the prevention of a chronic disease.

Reconceptualization of the RDA also needs to reconsider the stated applicability to 'practically all healthy persons'. After age 45, most Americans are not 'healthy' in the strict sense of the word [DHHS/USDA, 1989; Lachance, 1992]. Further, the proposed expansion of separate RDAs for the elderly is not contested. Elderly have increased needs for several vitamins (B_2, B_6, B_{12}, D) as well as folic acid [Russell and Suter, 1993; Selhub et al., 1993]. Dividing the RDAs for ages 51–69 and 70 and over is realistic; however, there

Table 1. Recommended diet: health factors of USDA/NCI recommended dietaries

Dietary health factor	USDA[1]	HHS (NCI)[2]	RDA (1989) adult
Calories	1,695	1,604	>1,520
Protein, g	84±8	84±5	50–63
Total fat, g	59±6	52±6	not specified
% calories from fat	31	30	30
Polyunsaturated fat, g	15±4	12±4	not specified
% calories from PUFA	8	6.7	not specified
Saturated fat, g	19±4	17±4	not specified
% calories from saturated fat	10	9.5	<10% of calories
P/S ratio	0.8	0.8	not specified
Cholesterol, mg	238±97	188±33	<300
Total carbohydrate, g	216±15	212±12	>200
Dietary fiber	28±	30±	not specified
Total vitamin A activity, RE (μg)	9,689	11,183	800–1,000
Preformed vitamin A, (IU)	919	1,018	not specified
Provitamin A (carotene), mg	5.2	6.0	not specified
% provitamin A (carotene)	90.5	90.9	not specified
Vitamin E (total)	27	23	8–10
Vitamin C, mg	225	217	60
Thiamin (B_1), mg	1.7	1.6	1.1–1.5
Riboflavin (B_2), mg	1.9	1.8	1.3–1.7
Niacin (B_3), mg	24	24	15–19
Vitamin B_6, mg	1.4	1.3	1.6–2.0
Vitamin B_{12}, mg	3.2	2.9	2.0
Folic acid, mg	353	381	180–200
Calcium, mg	1,004	1,017	800
Phosphorus, mg	1,371	1,420	800
Sodium, mg	1,887	1,955	>500
Potassium, mg	3,464	3,480	>2,000
Magnesium, mg	362	388	280–350
Iron, mg	14	14	10–15
Zinc, mg	13	13	12–15

[1] US Department of Agriculture.

[2] US Department of Health and Human Services, National Cancer Institute, National Institutes of Health.

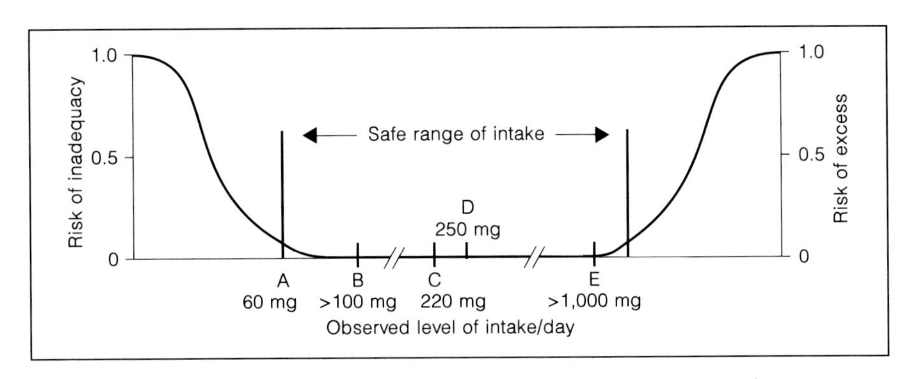

Fig. 1. A safe range of ascorbic acid intakes. The concept assumes that there is individual variability of both requirement for a nutrient and tolerance for high intake. The safe range of intakes is associated with a very low probability of either inadequacy or excess for an individual (RDI for Canadians). A=RDA (1989); B=RDA (smokers) (1989); C=USDA/NCI menus [Lachance, 1992, ACS No. 484, chapt. 26); D=for saturation [Simon: J Am Coll Nutr, 1992;11:107]; E=interference with other clinical test results.

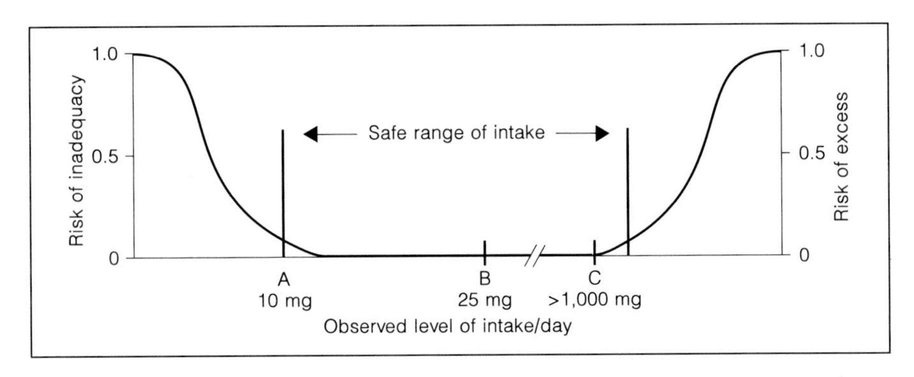

Fig. 2. A safe range of vitamin E intakes. The concept assumes that there is individual variability of both requirement for a nutrient and tolerance for high intake. The safe range of intakes is associated with a very low probability of either inadequacy or excess for an individual (RDI for Canadians). A=RDA (1989); B=USDA/NCI menus [Lachance, 1992, ACS No. 484, chapt. 26); C=Campbell et al. [1980], LSRO (FASEB).

is no apparent manner to adjust RDAs for assuring enhanced quality-of-life factors.

The timeliness of the RDAs could occur if the updating of the allowances were a continuous process. Each nutrient would be reconsidered as frequently as scientific developments dictate. A comprehensive review of all other nutrients should not be needed because a nutrient such as folic acid needed reconsidera-

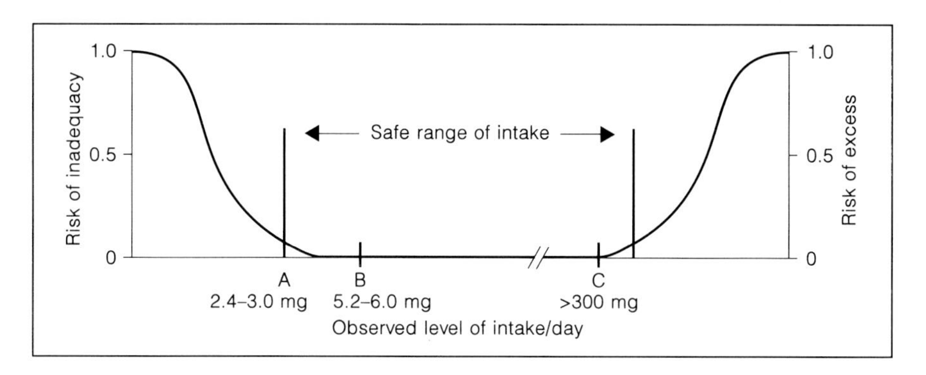

Fig. 3. A safe range of carotene intakes. The concept assumes that there is individual variability of both requirement for a nutrient and tolerance for high intake. The safe range of intakes is associated with a very low probability of either inadequacy or excess for an individual (RDI for Canadians). A = 50% of RDA (1989) for vitamin A; B = USDA/NCI menus [Lachance, 1992, ACS No. 484, chapt. 26); C = dosage/day for erythropoietic porphyria.

tion. This process would permit maximum outside input into the scientific deliberations. Consistency with other guidelines and the RDAs of other countries might be more easily considered. Further, conflicts between RDAs and standards established by other US regulatory agencies need to be reconciled to avoid creating unnecessary confusion. This would lead to wiser risk assessment philosophies in Congress and hopefully, with the American public. For example, trace minerals such as zinc, iron, nickel, chromium, selenium and manganese appear to be needed for nutrition at levels that toxicologists fear may be adverse [Abernathy et al., 1993]. Toxicologists might also learn to change the current method of determining safety. To relegate the responsibility of science policy directions to regulatory bodies would be a major mistake because regulatory bodies are users of scientific standards in a legal precedence sense (and therefore often outdated sense) and such bodies are not innovators nor educators of scientific advances. In this regard and for the prior reason, the RDAs must incorporate not only established nutrients and their applicable precursors, but must also examine and give consideration to food components that are not classified as 'essential' nutrients but which have physiological and/or biochemical implications in health such as soluble and insoluble fiber, dietary oxidized cholesterol and fatty acids, trans fatty acids, carotenoids, naturally occurring toxicants and antitoxicants and compounds affecting the metabolism of nutrients and toxicants.

In summary, there exists exceedingly strong evidence indicating a protective effect is conferred by the combination of water-soluble and lipid-soluble antioxidants with the resultant thwarting of a number of chronic diseases stemming from repeated oxidative insults. Confidence for dietary intervention recommendations is enhanced when epidemiological data is sustained by concomitant blood (tissue) values for the antioxidants. The establishment of desirable blood values from the composite experience of the WHO/Monica, Basel and VERA studies makes possible the determination of minimal antioxidant intakes to assure desirable blood levels. It is encouraging to note from data of this meeting that results from various types of analyses lead to a convergence of recommended intakes that can be made. Further, the adoption of a philosophy of safe and usual intakes, spanning requirements to thwart deficiency, to nutrient levels in 'ideal diets' which have been associated with chronic disease prevention, as well as clinical, intervention and metabolic study results, and to upper limits just before safety or functional effects occur, permits the establishment of RDAs realistically approaching desirable optimal intakes.

References

Abernathy CO, Cantilli R, Du JT, Levander OA: Essentiality versus toxicity: Some considerations in the risk assessment of essential trace elements; in Saxena J (ed): Hazard Assessment of Chemicals. Washington, Taylor & Francis, 1993, vol 8, pp 81–113.

Anonymous: Estimated Savings in Preventable US Hospitalization Expenditures with Increased Antioxidant Intakes: An Economic Analysis. Washington, Pracon Inc, 1993.

Block G: Dietary guidelines and the results of food consumption surveys. Am J Clin Nutr 1991;53S: 356–357.

Centers for Disease Control: Recommendations for the use of folic acid to reduce the number of cases of spina bifida and other neural tube defects. MMWR 1992;41(RR 14):1–7.

Centers for Disease Control: Recommendations for use of folic acid to reduce the number of spina bifida cases and other neural tube defects. JAMA 1993;269:1223–1228.

DHHS: The Surgeon General's Report on Nutrition and Health. Washington, US Government Printing Office, 1988.

DHHS/USDA: Nutrition Monitoring in the United States – An Update Report on Nutrition Monitoring. DHHS Publ No (PHS) 89-1255, Washington 1989.

Enstrom JE, Kamin LE, Klein MA: Vitamin C intake and mortality among a sample of the United States population. Epidemiology 1992;3:194–202.

Esterbauer H, Gebicki J, Puhl H, Jurgens G: The role of lipid peroxidation and antioxidants in oxidative modification of LDL. Free Radic Biol Med 1992;13:341–390.

Gaby SK, Singh VN: β-carotene, in Gaby SK, Bendich A, Singh VN, Machlin LJ (eds): Vitamin Intake and Health: A Scientific Review. New York, Dekker, 1991, pp 29–58.

Health and Welfare Canada. Nutrition Recommendations. Ottawa, Health and Welfare Canada, 1990.

Heinonen OP, Huttunen JK, Albanes D, et al: The effect of vitamin E and beta-carotene on the incidence of lung cancer and other cancers in male smokers. N Engl J Med 1994;330:1029–1035.

Hofstetter A, Schultz Y, Jequier E, Wahren J: Increased 24-hour expenditure in cigarette smokers. N Engl J Med 1986;314:79–82.

Jialal I, Grundy SM: Effect of dietary supplementation with alpha-tocopherol on the oxidative function of low density lipoprotein. J Lipid Res 1992;33:899–906.

Jialal I, Grundy SM: Effect of combined supplementation with alpha-tocopherol, ascorbate, and beta-carotene on low-density lipoprotein oxidation. Circulation 1993;88:2780–2786.

Kallner AB, Hartman D, Hornig DH: On the requirements of ascorbic acid in man: Steady state turnover and body pool in smokers. Am J Clin Nutr 1981;34:530.

Kant AK, Schatzkin A, Harris TB, Ziegler RG, Block G: Dietary diversity and subsequent mortality in the First National Health and Nutrition Examination Survey epidemiologic follow-up study. Am J Clin Nutr 1993;57:434–440.
Lachance P: Nutritional responsibilities of food companies in the next century. Food Technol 1989;43: 144–150.
Lachance P: Diet-health relationship; in Finely JW, Robinson SF, Armstrong DJ (eds): Food Safety Assessment. ACS Symp Ser No 484. Washington, American Chemical Society, 1992, pp 278–296.
Lachance P: Micronutrients in cancer prevention; in Huang M-T, Osawa T, Ho C-T, Rosen RT (eds): Food Phytochemicals for Cancer Prevention. I. ACS Symp Ser No 546. Washington, American Chemical Society, 1994a, pp 49–64.
Lachance PA: Scientific status summary: Human obesity. Food Technol 1994b;48:127–138.
Lachance PA, Langseth L (eds): Proceedings of a Workshop on the Future of the Recommended Dietary Allowances. New Brunswick, Cook College Continuing Education, Rutgers University, 1994.
NIH: Consensus conference on osteoporosis. JAMA 1984;252:799–802.
NAS/NRC: Recommended Dietary Allowances. Reprint and Circular Series No 115. Washington, National Research Council, 1943.
NAS/NRC: Proposed Fortification Policy for Cereal Grain Products. Food and Nutrition Board. Washington, National Academy Press, 1974.
NAS/NRC: Diet and Health. Implications for Reducing Chronic Disease Risk. Washington, National Academy Press, 1989a.
NAS/NRC: Recommended Dietary Allowances, ed 10. Washington, National Academy Press, 1989b.
Reaven PD, Khouw A, Beltz WF, Parthasarathy S, Witztum JL: Effect of dietary antioxidant combinations in humans: Protection of LDL by vitamin E but not by β-carotene. Atheroscler Thromb 1993;13: 590–600.
Rimm EB, Stampfer MJ, Ascherio A, Giovannucci E, Colditz GA, Willett WC: Vitamin E consumption and the risk of coronary heart disease in men. N Engl J Med 1993;328:1450–1456.
Rosenberg JH: Folic acid and neural-tube defects – time for action? N Engl J Med 1992;327:1875–1877.
Russell RM, Suter PM: Vitamin requirements of elderly people: An update. Am J Clin Nutr 1993;58: 4–14.
Selhub J, Jacques PF, Wilson PWF, Rush D, Rosenberg IH: Vitamin status and intake as primary determinants of homocysteinemia in an elderly population. JAMA 1993;270:2693–2698.
Stampfer MJ, Hennekens CH, Manson JE, Colditz GA, Rosner B, Willett WC: Vitamin E consumption and the risk of coronary disease in women. N Engl J Med 1993;328:1444–1449.
USDA/DHHS: Nutrition and Your Health: Dietary Guidelines for Americans, ed 3. Washington, Government Printing Office, 1990.
Weisburger JH: Nutritional approach to cancer prevention with emphasis on vitamins, antioxidants and carotenoids. Am J Clin Nutr 1991;53S:226–237.

Paul A. Lachance, PhD, DSc, FACN, Department of Food Science, Rutgers, State University of New Jersey, New Brunswick, NJ 08903-0231 (USA)

Walter P (ed): The Scientific Basis for Vitamin Intake in Human Nutrition.
Bibl Nutr Dieta. Basel, Karger, 1995, No 52, pp 168–172

Discussions in Working Groups

P. Walter

Swiss Vitamin Institute and Department of Biochemistry, University of Basel, Switzerland

In three ad hoc working groups the participants of the meeting discussed several questions related to the prevention of chronic disease and to some general aspects for future recommendations. In the final general discussion, the conclusions of each of the groups were presented and open questions were formulated. In the following the main points are summarized:

Prevention of Chronic Disease

Which vitamins are involved? Can the preventive potential for each disease be related to a single vitamin or to a combination of vitamins?

It was agreed that the antioxidant vitamins very likely play a role in the reduction of risk of chronic disease, although their exact role cannot be clearly defined as yet. There was also a general consensus that the preventive potential is probably the combined effect of several antioxidant compounds including vitamin C, E and β-carotene.

A lot of evidence has accumulated in the field of cardiovascular disease and for some type of cancers. Concerning cancer, it can almost be considered proven that a diet high in fruits and vegetables has a protective effect. However, the importance of the exact role of antioxidants as a whole and specifically of the antioxidant vitamins is difficult to evaluate at the present time. It would also appear that it depends at what stage of cancer formation the antioxidants become effective, since the Finland study could be interpreted that an intervention at a possibly late stage of cancer may even promote its development.

Regarding cardiovascular disease, the primary risk factor remains high energy intake, specifically the high percentage in dietary fat. The epidemiological evidence available so far support very strongly a preventive role of antioxidative vitamins; however, more results from well-designed intervention studies are still necessary. It is apparent from molecular studies how antioxidant vitamins may protect radicals generated as a result of lipid peroxidation. Special emphasis was also given to the fact that the antioxidant vitamins can overcome some of the negative consequences of smoking. It was furthermore mentioned that the metabolism of homocysteine, another risk factor of cardiovascular disease and its relation to folic acid, vitamins B_6 and B_{12} should also be considered; however, more results are needed to evaluate the importance of homocysteine metabolism for cardiovascular disease.

Concerning other diseases, the special preventive action of calcium together with vitamin D and maybe vitamin K, currently being investigated, were mentioned. More and more evidence is also available on the relation between carotenoids and risk reduction of cataract formation, whereas for Alzheimer disease and senile dementia the relationship with nutrients was considered to be not clear yet. Also renal failure is being more and more discussed in relation to high protein diets, and it is open whether radicals and a possible protection by antioxidant substances are important.

Iron deficiency was also a topic since a diet high in vitamin C increases non-heme iron absorption. Considerable emphasis, however, is given today also to the question whether vitamin C may result in an iron overload leading to an increased risk for more radical formation and possibly radical damage. This aspect seems to be especially important in women after their menopause.

Finally, recent evidence points to the involvement of radicals and the potential protective effects of antioxidant substances in cases of diminished immunocompetence as observed often in elderly people. Here again, more research is needed to establish whether vitamins can improve immunocompetence which in turn would of course strengthen the general health of the elderly.

The relationship between folic acid and its preventive potential in reducing the incidence of neural tube defect was discussed in detail. Since the critical time is already about 4 weeks after conception, the best prevention would be if all women in the childbearing age would have an intake of 0.4 mg folic acid/day. It was considered impossible to obtain this amount of folic acid with a regular diet and therefore special supplementation or food enrichment with folic acid is necessary. Even though many different opinions were expressed, a majority seemed to favor the enrichment of flour and cereals with folate.

Can the preventive potential for each disease be related to plasma concentrations of certain vitamins and what is known about the relation between intake and plasma concentrations of the various vitamins?

There is substantial evidence on the intake of vitamins and its relation to plasma concentrations; however, the kinetics for each vitamin are different. Some depend on a long-time intake, others respond very rapidly to changes in the diet. Furthermore, individual variation is quite large and for some vitamins it is therefore necessary to analyze multiple samples for each person. There are still also analytical problems, mostly concerning the stability of vitamin C, making comparisons rather difficult especially with results of the older literature.

Measurements of plasma concentrations of vitamins and their correlation with the incidence of a disease will be an important tool for the future. There was a general agreement that at the moment there is not enough evidence of sufficiently controlled studies to allow meaningful and really conclusive correlations. It was also stressed that it would be useful to have a more general functional test for the antioxidant status which would be less dependent on the immediate intake of the nutrients with food. This last point was emphasized very strongly and came up in the discussions many times. Of course, research on other antioxidants (e.g. bioflavonoids and other carotenoids) as well as with other food components that may play a role in prevention or risk reduction should be continued.

General Aspects

Should the concept of RDAs in its 'traditional' importance be maintained in view of their role in preventing vitamin deficiency and malnutrition?

This question was intensively discussed and many aspects were considered. A general consensus was achieved that the present RDAs have a very important established role. They are simply necessary because governments need a set of criteria for a 'minimal' food provision. They are also a standard for addition of vitamins in food enrichment and therefore ultimately for labeling. Even though the labeling policies are a very interesting and controversial topic, there was no time to discuss this aspect in further detail. If changes in RDA values for vitamins were to be made to include their effects on marginal deficiencies as well as their potential for the reduction of the risk of chronic diseases, then it would be necessary to have a substantial scientific agreement on broad and consistent evidence. It was felt that such solid evidence – maybe with the exception of folate – at the present time is not available and can especially not be related to single vitamins. It will also be very difficult to define exact amounts of single vitamins for any preventive action of chronic diseases.

Based on this discussion, there was a general support for the view that an alternate concept other than the present RDAs is to be developed. The possibility should be considered to maintain the present RDAs (prevention of avitaminosis) but to issue in addition some kind of new recommendations or guidelines addressed to the individual being interested in 'optimal health' on the basis of available knowledge. The media report a lot on newly emerging science in the vitamin area and the consumer would like to have some sound information giving the opportunity to decide on the vitamin uptake. The possibility to have some recommended ranges for groups of vitamins related to the risk reducing potential for certain diseases was therefore emphasized. The difficulty then, of course, would arise that one suddenly would have to deal with two sets of recommended values, on the one hand the classical RDAs and on the other hand some new, yet to be named recommendations or guidelines. This question, however, ought to be resolved because the consumer in the long run will not be satisfied with today's recommendations for the prevention of avitaminosis being no longer of real concern in the Western world. Future discussions are therefore needed to decide in which direction our recommendations should go, either (a) change in the present RDAs or more likely (b) leaving the present RDAs and formulating a new set of recommendations possibly in form of guidelines for the consumer; certainly such guidelines must also include safety considerations and it should furthermore be possible that such guidelines can be adapted frequently according to the newest scientific knowledge.

What are the consequences with regard to public health and policy statements?

The currently Acting Director of the Food and Nutrition Board of the Institute of Medicine (USA) presented very briefly the just issued brochure [1]: 'How should the recommended dietary allowances be revised?' It seems appropriate to end this book by citing a chapter out of this brochure summarizing the problems we have to cope with when we either change the present RDAs or when we issue additional guidelines on the newer developments for the use of vitamins:

'If reduction of risk of chronic disease is to become a criterion in the development of future RDAs, many questions must be faced. Among them are central questions about what the RDAs are meant to be: Are they levels of intake based on requirements for specific biochemical functions? Are they based on less specific physiological outcomes possibly related to multiple functions? If the answer is 'yes' to both, then it is possible and may be desirable to provide multiple recommendations based on different functional endpoints. Additional questions include the following: What criteria should be used to set recommended levels of intake when clinical trial data are lacking? What is the desirable level of intake over a lifetime?

How can desirable levels of intake be extrapolated for groups not included in clinical trials (such as children, adolescents, young adults, and the elderly)? Should levels of nutrient intake be expressed in terms of numerical ranges, in terms of food patterns, or in some other way? Under what conditions do the functions of nutrients consumed at levels above the amounts obtainable from food become pharmacological agents outside the domain of the RDAs? How can concerns regarding potential interactions among nutrients be addressed?'

Reference

1 Food and Nutrition Board, Institute of Medicine: How Should the Recommended Dietary Allowances Be Revised? Washington, National Academy Press, 1994.

Prof. Paul Walter, Swiss Vitamin Institute and Department of Biochemistry, University of Basel, Vesalgasse 1, CH–4051 Basel (Switzerland)

Subject Index